ENERGY R&D

PROBLEMS AND PERSPECTIVES

ORGANISATION FOR ECONOMIC CO-OPERATION AND DEVELOPMENT

The Organisation for Economic Co-operation and Development (OECD) was set up under a Convention signed in Paris on 14th December, 1960, which provides that the OECD shall promote policies designed :

- *to achieve the highest sustainable economic growth and employment and a rising standard of living in Member countries, while maintaining financial stability, and thus to contribute to the development of the world economy;*
- *to contribute to sound economic expansion in Member as well as non-member countries in the process of economic development;*
- *to contribute to the expansion of world trade on a multilateral, non-discriminatory basis in accordance with international obligations.*

The Members of OECD are Australia, Austria, Belgium, Canada, Denmark, Finland, France, the Federal Republic of Germany, Greece, Iceland, Ireland, Italy, Japan, Luxembourg, the Netherlands, New Zealand, Norway, Portugal, Spain, Sweden, Switzerland, Turkey, the United Kingdom and the United States.

***

TABLE OF CONTENTS

Part III

ENERGY R & D POLICIES IN OECD MEMBER COUNTRIES

FOREWORD

This report constitutes the contribution of the Committee for Scientific and Technological Policy to the Long Term Energy Assessment carried out by the OECD in response to a directive of the OECD Council and published in mid-January 1975 on the responsibility of the Secretary-General under the title of Energy Prospects to 1985. The report of the Committee has contributed to the analyses and discussion of the possibilities afforded by scientific and technological research contained in the Assessment. Nevertheless, since the present report attempts to provide a detailed analysis of the specific problems involved in the formulation and implementation of a scientific and technological research policy covering all aspects of energy, it was considered that it is of particular interest as an independent document and should therefore be published separately.

The sudden end of a low-cost energy situation has brought tremendous problems to the industrialised, as well as to those developing countries which at present lack hydrocarbon resources, not to mention the critical consequences for the international financial and monetary situation. The transition from a low-cost to a high-cost energy economy entails direct and indirect effects, the gravity of which, although perceived, is still difficult to evaluate. There is obviously a need to conserve energy as much as possible, and to develop alternative sources of energy which may liberate the world from over-dependence on limited or unreliable energy sources. In the past, a more intense energy research and development effort could have been conducted in the industrialised countries, but the availability of low-cost energy did not provide sufficient incentives for carrying out R & D at the largest possible scale. Now that we have the experience of the energy crisis, it is time to plan an adequate effort in this area.

The long lead-time of research and development must, however, be taken into due account in order to avoid the danger of unwarranted short-term expectations. Research and development on alternative energy sources is often very complex, and the implementation of a long-term strategy in this field will require actions which must be decided in the immediate future. Such a strategy should, in my opinion, be flexible. It should, for example, avoid that nuclear energy alone plays, within a few decades, the master role that petroleum has played during the last twenty years.

One of the main impressions emerging from this report is the great number and variety of scientific and technological opportunities which could increase old and new energy supplies and conserve energy. Therefore, the exceptional dangers which confront industrialised countries today should not blind them to the fact that their present dependence upon expensive and insecure oil supplies from one geographical region is a quite ephemeral predicament, if seen in historical perspective. However, this does not mean that, once this temporary, specific dependence is overcome, everything could or should fall back into pre-1973 conditions. The oil crisis has raised a number of issues which will remain with industrial societies, such as the need for a more rational, comprehensive and long-term management of resources including energy resources, and the need for understanding the interdependence between energy and the type of society or civilisation in which we choose to live.

Moreover, many of the changes which the oil crisis is about to provoke are likely to become permanent, or at least to last for much longer than this crisis. This applies to changes in industrial innovation, to the substitution of certain materials by others, to certain changes in the overall structure of industry, nationally and internationally. It also applies to the geo-political changes and the international balance of power changes which are likely to be brought about - changes which are very complex and which ultimately may not all be in favour of the major oil exporters of today.

Once again, if it depended on science and technology alone, there is little doubt that the industrialised world has the potential to deal effectively with today's energy crisis, and to make sure that the present deterioration of its economic strength does not turn into a continuous decline. However, it does not depend upon science and technology alone, and it is the interplay of economic and political factors, which are often less predictable than the technological ones, which explains some of the present disarray. Elucidating the interaction between R & D and these economic and political factors is a major aim of science policy analysis, and I believe this report helps to clarify this interaction.

To carry out this work the CSTP constituted an ad hoc Group which I have had the privilege to chair. The Directorate for Science, Technology and Industry (Directorate for Scientific Affairs until 1974) set up a small team, which visited a number of Member countries where discussions with experts from government, industry and university were held. A group of experts of international reputation acted in a consulting capacity, and met in Paris during the preparation of the report. At the same time, Member countries helped to carry out a survey of national energy R & D programmes, which has become the basis of Part III of this report.

The ad hoc Group met four times to discuss the report, but the burden of drafting, revising and completing it has been carried by the Secretariat staff to whom I wish to address my thanks. Also, I should like to gratefully acknowledge the contribution of the numerous experts and consultants of the various Member countries, as well as that of the government administrations involved.

Without the help from all these sources, I believe that it would not have been possible to produce a report which I hope will be stimulating and useful not only to OECD Governments, but to all those who are concerned with the scientific and technological implications of energy problems.

Umberto COLOMBO
Chairman of the
Committee for Scientific
and Technological Policy

ABBREVIATIONS

AEC = Atomic Energy Commission (United States)

ANVAR = Agence Nationale de Valorisation de la Recherche (France)

BGC = British Gas Corporation

BNFL = British Nuclear Fuels Limited

BRGM = Bureau de Recherche Géologiques et Minières (France)

CEA = Commissariat à l'Energie Atomique (France)

CEGB = Central Electric Generating Board (United Kingdom)

CNEXO = Centre National pour l'Exploitation des Océans (France)

CNRS = Centre National de la Recherche Scientifique (France)

COST = Coopération Européenne dans le Domaine de la Recherche Scientifique et Technique

CSIRO = Commonwealth Scientific and Industrial Research Organisation (Australia)

DGRST = Délégation Générale à la Recherche Scientifique et Technique (France)

DOD = Department of Defense (United States)

ECSC = European Community for Steel and Coal

EDF = Electricité de France

EMR = Department of Energy, Mines and Resources (Canada)

FOA = National Defence Research Institute (Sweden)

GDF = Gaz de France

GKSS = Gesellschaft für Kernenergieverwertung in Schiffbau und Schiffahrt mbH (Association for the Exploitation of Nuclear Energy in the Ship Building and Shipping Industry) (Germany)

HUD = Department of Housing and Urban Development (United States)

IAEA = International Atomic Energy Agency

IGS = The Institute of Geological Sciences (United Kingdom)

IFP = Institut Français du Pétrole

IIASA = International Institute for Applied Systems Analysis

IMCO = Intergovernmental Maritime Consultative Organisation

JAERI = Japan Atomic Research Institute

KEMA = Joint Research Institute of the Dutch Electric Supply Companies

KTH = Royal Institute of Technology (Sweden)

MITI = Ministry of International Trade and Industry (Japan)

NASA = National Aeronautics and Space Administration (United States)

NEA	= Nuclear Energy Agency (OECD)
NSF	= National Science Foundation (United States)
NATO (CCMS)	= NATO (Committee for the Challenge of Modern Society)
NCB	= National Coal Board (United Kingdom)
NTNF	= Royal Norwegian Council for Scientific and Industrial Research
ÖEF	= National Board of Economic Defence Planning (Sweden)
ORSTOM	= Office de la Recherche Scientifique et Technique Outre-mer (France)
PNC	= Power Reactor and Nuclear Fuel Development Corporation (Japan)
RCN	= Netherlands Reactor Centre
STU	= Swedish Board for Technical Development
TNO	= The Organisation for Industrial Research (Netherlands)
UNEP	= United Nations Environmental Programme
UKAEA	= United Kingdom Atomic Energy Authority
USGS	= U.S. Geological Survey
VEG	= Joint Research Institute of the Gas Companies (Netherlands)

INTRODUCTION

This analysis of the problems and prospects of scientific and technological research in the field of energy has been drawn up by the ad hoc Group set up by the Committee for Scientific and Technological Policy in May 1973 under the chairmanship of Dr. Colombo in order to prepare the Committee's contribution to the Long Term Energy Assessment called for by a Council Resolution of October 1972.

During the period May 1973 to September 1974, the ad hoc Group held four meetings. From the outset, it took the view that its contribution to the Organisation's overall study should provide as detailed an analysis as possible of the problems involved in the formulation and implementation of a scientific and technological research policy covering all aspects of energy. After reviewing the medium- and long-term prospects of scientific and technological research, the Group therefore proceeded to gather information concerning Member countries' research and development at both national and international level, and to draw policy conclusions with respect to those sectors where a more intensive research effort was called for both within the framework of national policy and in support of international co-operative action.

Based on comments received from Member countries, an outline of the information to be requested and of the different problems to be tackled, was drawn up by September 1973 and by the end of February 1974 most countries had replied to the Secretariat. The Secretariat also visited certain countries for discussions with energy R & D experts in national administrations, universities, government laboratories and industry, notably in Canada, France, Germany, Japan, the Netherlands, Sweden the United Kingdom and the United States. Many individual experts and groups of experts were consulted with the object of assessing scientific and technological trends and potential in the various sectors of energy research. The results of this part of the survey were discussed on 21st and 22nd March, 1974 at a meeting of experts (see Annex III).

Part I deals with the fundamental role of energy in the development of society, its links with other natural resources and with the environment as a whole, and finally the international dimension of all aspects of energy problems, particularly R & D.

Part II attempts an assessment of the possibilities offered by scientific and technological research, and the obstacles it encounters, covering all existing and potential energy sectors and all energy applications; special attention is paid to the development of technologies for rationalising the production, utilisation and conservation of energy resources and for exploiting the possibilities of new sources of energy.

Part III consists of an analysis of R & D in Member countries; stress is laid on problems of planning and finance, institutional aspects, and the formulation of a more coherent research policy. This Part also includes a review of various sectors in which co-operation appears desirable, if not essential, to accelerate research results and minimise costs. Consideration is also given to ways in which R & D in industrialised countries could contribute to the solution of energy problems in developing countries, which have been greatly accentuated by the sharp rise in the price of oil. Finally, an attempt has been made to identify priority sectors for co-operation where a joint solution to the scientific and technological problems of exploiting both conventional and new energy sources would be more effective and less costly.

In accordance with the general tenor and conclusions of the ad hoc Group's discussions, which were communicated to, and approved by, the Committee, energy R & D problems and trends were catalogued and analysed, from the outset, within the broader framework of overall scientific and technological policy, the purpose being not merely to deal with short-term objectives, but also, and especially, to emphasize that medium- and long-term energy R & D programmes, whether national or international, must inevitably be based on systematic and overall consideration of all the technical, institutional and social problems involved.

The Secretariat would like to take this opportunity to thank all those in government, in the universities and in industry in the different countries whose views and advice have helped it to plan and conduct the survey on which this report is based.

Part I

GENERAL REMARKS

One of the consequences of the 1973 oil shortages was an increased awareness of the benefit to be had from intensified research and development, not only in finding, exploiting and bringing into use further energy resources, and in decreasing dependence on oil, but also in reducing the rate at which the demand for energy was growing whilst using present day technologies. The crisis has also spotlighted the weaknesses of earlier policies: while there had been a sustained research effort in certain sectors, e.g. nuclear energy, others have been neglected and certain aspects, in particular a more rational use of energy, were almost completely disregarded.

Research and development is one means of promoting the achievement of objectives set by an overall energy policy, and the size and content of R & D strategies should not be decided independently of that policy. In view of the time needed to implement research programmes and then to apply the results, this is a relatively urgent task, but one which needs to be correctly formulated before expensive R & D is commenced.

Energy policies must clearly be based on the prospects offered by the various technologies, on the extent and distribution of the different energy resources, and on the economic and industrial structure as well as the scientific and technological capacity of each country. It is evident that any medium- and long-term energy policies or objectives must include R & D efforts as a key component. It is therefore important before examining the main questions involved in energy R & D, concerning both research content and structure, to indicate some basic features of energy, and to see what effects they have, particularly as regards science and technology.

In particular, although this report is intended to cover neither the management of natural resources as a whole nor the sociology of energy, not even the geopolitics of energy, it is important to set out a number of self-evident but basic considerations concerning the role of energy in society, its links with other natural resources and its international character. In fact these data form the general framework within which any coherent scientific and technological policy for energy must be expressed and implemented.

1. The place of energy in society

It is clear, without going into such detail as the fundamental equivalence of matter and energy, or the energy aspects of the biological phenomenon of life, that all man's needs and activities, both essential and non-essential, depend on energy. This dependence may be direct and physical as in the case of transport or heating, or indirect where energy is necessary for obtaining and converting into finished products the material resources which are also essential to any human activity. The converse, although usually forgotten, is no less true and no less important: energy has real importance only to the extent that it is used; energy has no value in itself - only in the function that becomes possible through its use.

Any society depends on energy if only because energy is needed to produce food, but the ways in which energy is obtained and used have a far-reaching influence on the forms of civilisation, life styles and social structures. For example, referring again to food, the "invention" of agriculture, in other words the organised use of solar energy processed by plants, was a decisive factor in the trend towards nomad settlement which was a feature of the neolithic era. More recently, the use of coal in the steam engine and oil products in the internal combustion engine and the harnessing of electrical power profoundly affected both the economy and the structure and development of industrialised societies. The form that economic growth has taken in these societies over the last two centuries has been marked by the growth in energy consumption to such an extent that the two trends have come to be regarded as inseparable. Conversely the type of civilisation and structure of a particular society have a definite influence, through science and technology in particular, on the nature of the energy resources that it uses and the way it uses them. An example is the harnessing of nuclear energy which is a direct result of the scientific and technological capacity of industrialised society of today.

Two conclusions may be drawn concerning research and development from these two elementary observations, namely the permanent need for energy and the influence of energy on society and vice versa. Firstly, energy should be the subject of constant vigilance, even if its importance is really recognised only in times of crisis. Because of the time scale of the process of research and of the application of its results, the goal of R & D should be primarily the continuous satisfaction of medium- and long-term energy requirements. Although it is true that the present situation has made it imperative to think again about energy matters and to review the R & D effort, there is also a risk that too much weight will be attached to short-term considerations. In other words, one of the roles of research is not so much to solve the problems of the present crisis as to prevent as far as possible the occurrence of others in the future and to provide the means of dealing with those which cannot be avoided.

The second conclusion is that energy R & D policies should not be limited to the production of energy, but should also cover the way in which it is used and the effects of the energy production and consumption systems on the structure of the economy and society. Technological decisions are less neutral in the field of energy than in any other field; they must be considered in relation to political options affecting the future development of society.

2. Energy and other natural resources

Energy is not an "independent" resource and its function in human society can be fully understood only in the light of its many and complex links with the other kinds of natural resource on which society depends, particularly raw materials and the environment regarded not solely from the pollution standpoint but more generally in terms of resources such as water, air and land.

With regard to energy and materials, a first point is that certain natural resources may provide both energy and materials; one example is petroleum which could also become a substantial resource of synthetic food protein. A second point is that energy is consumed in obtaining, converting, using and recycling material resources; as lower quality deposits are exploited, more energy has to be used to win a given quantity of usable material; substituting other materials in a given application can appreciably alter the amounts of energy needed for this purpose; lengthening the lifetime of materials and products generally reduces the associated energy demand; organic wastes may constitute a source of energy. Lastly, energy cannot be produced, converted, transported, stored or used without materials; Part II of this report, which deals with future possibilities in science and technology, shows how critically the development of all energy technologies is dependent on the development of suitable materials, throughout the energy cycle - from production to consumption.

The relationships between energy and the environment are equally complex. Firstly there are those types of pollution that are directly connected with the various energy sources, e.g. carbon dioxide emitted by fossil fuels, or the thermal pollution associated with any consumption of energy. Any reduction in the amount of energy brings about a reduction in total pollution in this form but it must be remembered, that the geographical pattern of pollution is at least as important as its total amount. Moreover, there is at present no satisfactory way of comparing the different kinds of pollution; for example, how can the carbon dioxide emissions from fossil fuels be compared with the radiations from nuclear power generation?

These problems are even more acute for those kinds of pollution (such as that from chemical plants) which do not stem directly from the production or the consumption of energy. When several technologies fulfil the same function, the best one from the environmental conservation standpoint is not necessarily that which consumes the least energy. In addition, the act of monitoring, conserving or restoring the environment may itself involve the consumption of a significant amount of energy; a case in point is the energy used in the landscaping work required to make good the ravages of open-cast coal mining or other workings. In general, it can be stated that reduction of pollutants at their source is less expensive to society and less energy-consuming than their abatement when they are dispersed in the environment.

Lastly, the link between energy and environment has to be viewed in a broader context than that of pollution and environmental degradation. The production, conversion, transport and storage of energy require not merely energy resources "per se" but also other resources such as land or water; the supply of these is limited, variable in quality, unevenly distributed and they are also needed for other human activities with which energy is therefore in competition. Cases in point are the use of water for irrigation and energy production, the problems of land use involved in open-cast coal mining, power transmission lines and the siting of power stations and oil refineries.

It is therefore clear that R & D policies in the energy field must take full account of the many and complex relationships through which energy is associated with material resources and the environment in the widest sense of the term. Ideally, these policies should be an integral part of a general strategy covering all natural resources, the relationships between the various types of resource and their utilisation by society. However this is no good reason for waiting for some hypothetical date when such a strategy might be formulated before framing and applying R & D policies in the field of energy.

3. International aspects

In all aspects of the field of energy - resources, requirements and technology - there are profound differences between countries. These differences themselves imply a considerable measure of interdependence between the countries, and R & D policies in the energy field will therefore have to take into account this duality: the specific nature of individual countries' problems and the international character of the energy question.

As regards resources, the differences are, of course, a matter of their unequal geographical distribution in quantity and quality from country to country. This inequality is particularly striking in the case of oil but it also affects all other energy resources, including the sun and geothermal energy. In addition, the importance that off-shore oil and natural gas seem likely to assume is an intrinsically international problem of increasing acuteness which is still far from being resolved.

Differences in the energy requirements between countries are of a similar magnitude. They depend in particular on the country's economic growth, industrial structure and life style. Another important point is that energy requirements also vary widely in qualitative terms, even if only because of geographical differences. For example, the energy systems which best suit a small country are not necessarily the most appropriate for a country extending over thousands of kilometres, even though energy consumption per head of population may be the same in both.

Lastly, differences in scientific, technological and industrial potential from country to country also have a fundamental bearing on the question. For instance, the accelerated development of nuclear power envisaged by many OECD countries during the 1973 crisis is hardly conceivable in the developing countries even though they were equally affected by the rise in oil prices.

Whilst national R & D policies have to be based on the specific conditions applicable to each country, it is because of these differences and the interdependence of all countries that such policies must take into account the international character of the energy problem. Energy raises trans-national problems and countries have to co-operate. Furthermore, co-operation is necessary not merely to avoid unnecessary duplication of effort but also because of the scale of the problems, the rising cost of research and the complexity of the technologies involved.

Part II

SCIENTIFIC AND TECHNOLOGICAL PROSPECTS

INTRODUCTION

The following chapters constitute an evaluation of the perspectives which R & D are opening up in the energy field. This evaluation endeavours to create a basis for national science and technology policies relevant to all energy questions. It was therefore necessary to present as comprehensive a view as possible, including not only the exploration of many different energy resources, but also the conversion, storage, transportation and utilisation of different types of energy, as well as the relations between the different components of the overall energy system.

Apart from exclusively scientific and technical data, this evaluation touches upon the development lead-times and costs of various technologies, upon environment and safety problems, and finally upon the reactions of public opinion. Thus, an attempt has been made to pinpoint the main problems which R & D will have to tackle.

However, this evaluation is tentative and limited in several respects. First, there is an unpredictable element in all R & D; hence, present projections should not be considered as final, they will require continuous updating in response to new R & D results. Second, the wish to present the most comprehensive view of energy problems, did not permit the authors to review technological details in sufficient depth. Third, the present evaluation is not more than one of the factors upon which countries can base their energy R & D policies. In this connection, it is clear that each country's energy resources, research potential and industrial capacity, as well as many other geographical, political, economic and social factors, are of fundamental importance.

Inevitably, a comprehensive energy study raises classification problems. In face of the complexity of the energy problems and the great number of interdependent factors which they comprise, almost any classification is a simplification. Such a simplification cannot fully take into account all relevant technical, economic and institutional facts, nor does it allow for final judgements based on scientific rationality or systems analysis. The classification proposed in this paper is a compromise. The first three chapters on fossil fuels, nuclear energy and other energy sources, deal with the various primary energy sources and their conversion. The following chapters investigate energy carriers and the utilisation of energy - the aim has been to emphasize those components of the energy system which have hitherto been somewhat neglected in comparison to the problem of energy resources.

The variety of units of energy represents another serious difficulty in any report dealing with the whole field of energy. For each chapter, the units adopted have been those most commonly used in the relevant sector: e.g. barrel in the case of oil, kilowatt-hour (kwh) in the case of electricity. In order to facilitate comparisons and to ensure compatibility with the other OECD reports, the equivalents in ton oil equivalent (toe) as regards amounts of energy and dollar per barrel, as regards costs, have been indicated in footnotes. Moreover, as the joule is officially the international scientific unit of energy, conversions into joule and dollars per joule have also been made.(1) It must be added however that comparisons should not be made only on a quantitative basis but should also take full account of the qualitative differences between the various forms of energy: in practice, as well as in theory, one joule of thermal energy is not equivalent to one joule of mechanical or electrical energy.

1) For conversion factors for units of energy see Annex IV.

I FOSSIL FUELS

A. EXPLORATION AND ASSESSMENT OF FOSSIL FUEL RESOURCES

Up to now, data available regarding identified energy resources and reserves - mostly reserves of oil and coal, however defined - have been one of the chief factors upon which governments have based their energy policies. It appears that this basis has been and still is, rather unsound. It has been incomplete in most cases, and definitely wrong in some. Data sources vary widely not only between but within countries. While the private oil industry is in most countries the only source of comprehensive data on oil resources, national geological surveys have provided most information on coal. Certain resource definitions are so complex and subtle that, more often than not, they have led to misunderstandings, misquotations and finally miscalculations. In more than one case, resource data based on statistical estimates have been modified to serve political purposes. The borderlines between various categories or qualities of the same resource are not rigid; changes in price and technology modify them continuously. In a time of rapid change in both, for example in 1973 and 1974, almost the only certainty is that precise resource data are no longer valid the very moment they appear in print. Hence, one of the first tasks is the formulation of clearer, internationally comparable definitions and a faster and more comprehensive statistical updating of primary energy sources.

Energy R & D should start with a more systematic and continuous search for primary resources. On a world-wide basis, the distribution and quantities of on-shore oil and natural gas are relatively better known than those of other fossil fuel resources, which does not at all mean that they are well known. Global resources of coal, especially of lower quality coals, have probably been less well assessed than oil resources. Lastly, our knowledge of the geographical distribution and the extent of the reserves of tar sands and oil shale is most limited and fortuitous. A more accurate knowledge of recoverable reserves and of the distribution of each of these resources is urgently required. Such knowledge would alone make a decisive impact on present energy decisions.

Until approximately 20 years ago, prospection for fossil fuel resources was technically a relatively simple matter. The scientific and technological potential for much more sophisticated prospecting methods has recently increased substantially and this may help to place prospection on a much more sound scientific basis than was ever possible before. More R & D will be necessary to ensure the wider application and constant growth of available scientific and technological knowledge. In the oil sector for example, it will take much more geological, geophysical and geochemical research for a better understanding of the formation, migration and accumulation of oil and more advanced scientific techniques and instruments to ensure a reduction of the margin of error in oil exploration. If possible, the ultimate goal of exploration R & D should be the discovery of methods for the direct detection of sub-surface oil or gas. Traditional methods which can only detect the presence of geological structures favourable to oil or gas accumulations, are still far from this goal. Perhaps "Bright Spot" techniques will be among the first direct detection methods. These are advanced seismic data interpretation techniques which can apparently distinguish between seismic echoes from rocks, liquids and gas.

For the time being, spacecraft and aircraft remote sensing technologies are very promising innovations in the exploration of oil, coal and other resources. They could, for example, lead to a much better understanding of the sedimentary basin systems where oil can be found. NASA's Earth Resource Technology Satellite (ERTS) has already brought to light unknown geological features which might lead to the discovery of new mineral and oil fields. In this case, the technological innovation is not so much the satellite as the photographic and other equipment which it carries, and the most difficult and costly work is not the photographing and remote sensing itself, but the interpretation of the data. ERTS might to some extent make possible the prospection of oil, water and other resources on a world-wide basis, transcending national frontiers. Although the raw data obtained from ERTS are available to anyone and are in the public domain, the expensive process of analysing and interpreting these complex data could lead to results which are proprietary information. This could raise many problems related to economics, international relations and co-operation and which go beyond the scientific and technological aspects of ERTS. Of course, other methods such as improved seismic ones, might be just as important in the future for improving oil prospection.

R & D has to follow many paths in attempting to improve methods for finding oil, coal or other resources. Among many specific research opportunities, may be mentioned the use of large capacity computers for the world-wide co-ordination of petroleum geology data for the purpose of geophysical modelling, the development of more precise seismic and gravimetric methods, the application of isotope analytical techniques, etc.

B. THE OIL SECTOR

There are large sedimentary on-shore regions which could bear oil, but have not been prospected. Major discoveries are still possible in several continents. Apart from oil prospection, it is drilling which offers R & D a wide field for technological improvement, especially as far as drilling speed and depth are concerned. High-speed jet-drilling, which bores holes without mechanical contact, appears to be a very promising new technology. Other innovations which deserve further interest and support, are continuous drilling processes, and the utilisation of computers in oil drilling. With present drilling methods a depth of nearly 8,000 metres is reached but it is possible that new gas or even oil fields could be found at greater depths. However, the new fast drilling techniques should be devised and applied in such a way that the geological formations are not damaged.

A scientifically more advanced continuous interpretation of core samples, and other factors could considerably improve the judgement of oil drillers even before they strike oil. More R & D should be done to help drillers understand when and where they should continue and when and where they should give up drilling.

In addition to prospection and drilling, oil pumping technologies can also be improved. When natural gas, which is usually found together with oil flows into an oil pump, the latter often seizes up. Better methods of separating oil and gas in order to prevent this would be very useful.

The growth of the oil-trade and the discovery of oil and gas in remote or difficult regions, such as the Arctic, underline the importance of transportation technologies. Thus, the improvement of land and sea transportation of oil is an increasingly important target of R & D. The unpredictability of oil supplies has given a new, political urgency to the existing need for more oil storage facilities. There are several R & D programmes to increase and improve both oil and gas storage facilities. For example, high pressure storage, underground storage and storage on the sea-bed are being investigated, and more could probably be done in this sector.

1. Off-shore Oil

The shelf surrounding the continents, is said to be at least as promising as the continents themselves. The return on R & D investments could be considerable in continental shelf, or off-shore areas. The more optimistic experts believe that on a world-wide basis, there is more oil and gas to be found off-shore than all that has been, and still will be, discovered on-shore. However, off-shore oil makes new and different demands on science and technology.

Off-shore oil activities can be broken down into several stages: geological prospection, drilling, wellhead completion, production and evacuation or pumping to the shore. These are the most important ones. Of these activities, prospection is the only one which is easier off-shore than on-shore. The mobility of ships, and the absence of population, topographical, and other obstacles at sea, make off-shore prospection, especially with seismic methods, easier and cheaper.

However, off-shore drilling and production are technically much more challenging and expensive compared to traditional on-shore methods. If geological data indicate that there is a possibility of finding oil or gas under the ocean floor, an exploratory well has to be drilled. Drilling techniques vary according to geographical environment and ocean depth. Apart from very shallow waters, where artificial islands can be built, drilling is done from anchored "semi-submersibles", from anchored ships, from "jack-up" rigs which are floated to location and placed on legs, from huge fixed platforms, or, in greater water depth than 200 m., from dynamically positioned, mobile rigs. Ocean floor drilling has been technically possible at a depth of approximately 700-800 m. and seems to have become possible even at 1,000 m. as from April 1974, but the high costs and the limited number of available drilling ships and platforms are among the biggest technical bottlenecks in off-shore activities. Here R & D will help to reduce the costs and increase the supply of oil and gas. For example, the development of new floating rigs made of cheaper materials and dynamically stabilized could have a certain impact.

Experts are confident that further technological progress will make drilling possible at much greater depths; 2,000 m. under water has been mentioned as a target that could be reached in a not too distant future.

The recovery of oil from the ocean bottom is technologically very different from traditional on-shore activities. If the main problems of underwater oil production are soon solved, it will be thanks to a considerable R & D effort, to the application of advanced technologies, and to technology transfers from other industrial sectors.

If oil is found, the first step towards production is the installation of a "wellhead cellar". This is essentially a hull, for example a steel pressure vessel, which encapsulates the wellhead and the necessary valves, pumps or tools. It keeps the wellhead dry and permits service personnel to control and repair it in a dry atmosphere, except in more shallow waters where the same work could be done by divers.

Apart from wellhead cellars, an off-shore oil production system needs several other components. Most important is the "service capsule", a diving capsule which, like a diving bell, is lowered from a support boat and can bring a small number of workers down to the wellhead. In one of the most advanced off-shore systems, this service capsule is a direct technology transfer; it was originally developed to save sailors from sunken submarines.

The technical components for an off-shore system are available although certain problems related to oil storage on the ocean bed and pumping oil to the surface still have to be resolved. What remains to be done, is to assemble the technical components, and to compare the economics of competing systems. At least nine different off-shore systems have been proposed, and two are nearly ready for use. From 1974 on, oil production should, in principle, be possible up to a maximum water depth of 400 m.; in practice, full-scale off-shore production through new systems might start only in 1977, if present development is not accelerated. Later on, the technical possibility for descending to a depth of 3,000 m. exists, and the necessary hardware could be developed with an additional R & D effort.

Geological, or better geophysical, exploration reached a water-depth of 450 m. in 1973, will reach at least 650 m. in 1974 and may attain 2,000 m. in 1976. However, without corresponding technical advances and cost reductions in off-shore drilling and production especially at a depth of over 200 m., the time-lag between exploration and first production could remain quite long. Until 1974 it was approximately eight years. Of course, increased oil prices, government support or greater risk-taking by the oil industry could reduce this time-lag, but a reduction through R & D is certainly a more desirable solution.

An advance in ocean technology is all the more urgent in view of the fact that many oil-bearing off-shore wells had to be closed because the production technology was apparently not available. Moreover, geological studies have discovered numerous, huge sedimentary basins beyond the continental shelf, at the base of the continental slopes in waters between 2,000 and 3,000 m. deep. According to a recent hypothesis which is partly based on successful laboratory experiments, the conditions for the formation of oil in these basins are exceptionally good, and many of the world's richest oil fields could possibly be found there. The technology to test this hypothesis should be developed as quickly as possible.

Off-shore activities have added a new environmental dimension to oil production; the ocean environment makes considerable new demands on R & D. The danger of oil-spills has preoccupied governments and public opinion most, although the record so far indicates that this danger is technically very small and that blow-outs are nearly always due to human and not technical faults. During the last 20 years, tens of thousands of off-shore wells have led to no more than a small number of insignificant spills and a major one (Santa Barbara). On balance, the development of off-shore oil fields in relative proximity to consumption centres might damage the ocean environment less than the long-distance transport of the same amounts of oil by tankers. Although technologies to fight ocean spills have advanced considerably during the last few years, additional R & D might be necessary in this field, as well as in other fields related to the reliability of ocean oil technologies.

In Arctic areas, especially Arctic ocean areas, unique environment problems are often accompanied by unusual technical problems due to ice and cold.

On the whole, off-shore oil production will profit from, and in some cases will depend upon, further advances in all oceanological sciences. Certainly more R & D will be required to study climatic conditions, especially ocean currents, ocean floor geology, marine biology, corrosion of different materials in the ocean environment and many related subjects. Also, more R & D will have to be directed towards improving safety factors.

2. Secondary and Tertiary Recovery(1)

Technology advances are opening up new oil reserves which in the long run, will be vast: the unrecovered and hitherto unrecoverable parts of known oil deposits. In 1973, the proportion of oil per deposit which could be recovered averaged not more than 30 per cent. Experts believe that this proportion could probably be doubled in the long run although it is most unlikely that it will

1) The distinction between secondary and tertiary recovery is not everywhere the same. Usually, secondary recovery relates to oil obtained by the augmentation of reservoir energy, often by the injection of air, gas or water. Tertiary recovery means the use of heat and methods other than fluid injection to augment oil recovery, and takes place after secondary recovery.

ever reach 100 per cent in more than a very few cases. However, even reaching a 60 per cent recovery on average will require a considerable R & D effort and some time.

During the last few years, secondary and tertiary recovery research has had some effect. Although most oil is still being recovered without such methods, in the United States for example, fluid injection has already been responsible for about 35 per cent of total oil production in 1974. During the last few years the proportion of recoverable reserves increased by approximately a half per cent annually. During the coming ten years, higher oil prices and better technology will probably increase the recovery proportion by 1 per cent annually. In the United States alone, this would be equivalent to about 4 billion barrels annually, which is nearly equal to the total annual oil production of the United States in 1974. Thus, in the 1980s at least 40 per cent of the known oil in place should be recoverable, as compared to 30 per cent today, which alone will increase recoverable reserves by huge quantities.

One of the most widely used secondary recovery techniques is water-flooding, which includes pumping water down into oil-bearing rocks to force the oil out. More advanced techniques will use steam, heat and chemicals such as surface active agents. The latter will be added to the water to extract the oil which is hidden in countless very small rock cavities and ramifications. Until recently, the price of those chemicals and the quantities required were prohibitive.

As a result of additional research it should be possible to improve recovery methods. For example, a more precise knowledge of how oil and water are attached to the rock basis might be very useful; some experts are convinced that the geological and chemical knowledge of this subject is very incomplete. An idea which may be promising is to apply tertiary recovery methods from the very beginning, instead of producing the oil as long as possible with simple techniques. It seems that this could raise the ultimate recovery rate and reduce final costs per unit, more so than the traditional method of increasing the sophistication of recovery techniques only gradually and as necessary.

3. Heavy Oils

Heavy oils are not sufficiently fluid and, therefore, cannot be pumped to the surface without preceding treatment. For the time being, they are found mainly in Canada, the United States and

Venezuela. The extent of reserves is not well known because heavy oils have not raised much interest at oil prices of $1 or $2 per barrel(1).

Producing heavy oils in large quantities, which would certainly be economically advantageous in 1974, will require additional R & D. Recovery methods will sometimes be the same as, or similar to, the secondary and tertiary recovery methods for normal oil.

C. NATURAL GAS

Natural gas can be found together with oil ("associated gas fields") or independently from it ("unassociated gas fields"). Some gas has been generated from oil, some from coal. The exploration and production problems are largely the same as for oil, and resource data are not more reliable than those concerning oil.

However, in addition to the problems which it shares with oil, gas faces R & D with several specific challenges. Most important are storage and transport problems. Gas is the cleanest and most flexible natural fuel, but it is more difficult to store or to transport overseas than liquid and solid fuels. In unassociated gas fields, the production of gas can be regulated. In associated fields, it is linked to the production of oil and its flow cannot be modified separately from that of oil. Gas which is not used can be reinjected, destroyed or liquefied. Reinjection helps to maintain pressure in the oil-field and preserves the gas for future use. However, much natural gas which is produced together with oil, especially in the Middle East, is still destroyed through burning (flaring).

The most obvious solution to this wastage is to convert the gas into a fuel which can be easily transported and stored. The usual method today is to transport it by pipeline or in liquefied form, as "LNG", if it has to be transported overseas. However, LNG transport requires special, expensive ships and although this technology is quite mature, some experts have misgivings about the possibility of accidents which could release and ignite the LNG, especially in harbours. Though it may be increasingly difficult to attain new and better results, additional R & D should be devoted to the problem of cheap and simple liquefaction, transportation and storage techniques for natural gas, and to reducing the danger of accidents. LNG is perhaps not always the best solution; gas could also be converted into methanol for example. This is more expensive as half of the gas is lost by traditional conversion methods, but the transportation of Methanol is less dangerous, and

---

1) $1.7 or $3.4 per 10^{10} joule.

over long distances, cheaper than that of LNG. methanol is very valuable in the chemical industry, and could become an important transport fuel. Obviously, the future demand for methanol will influence decisions on natural gas conversion at least as much as transport technologies.

Off-shore gas, like off-shore oil, raises a number of very specific problems and thus creates additional challenges for R & D. Apart from problems related to the reliability and durability of the off-shore equipment, which are common to gas and oil, the re-injection of unused gas into wells on the ocean bed requires additional R & D. If the off-shore gas is neither reinjected nor brought ashore, an economic solution would be to liquefy it on platforms and move it by LNG tankers, or to convert it into methanol. More R & D is necessary for the study of cheap, reliable on-platform liquefaction and methanol conversion technologies.

Other tasks in the natural gas sector which still lie before us, are the evacuation and use of the gas which is sometimes hidden in coal seams, and releasing the gas bound in "tight rock" formations. In the United States alone, gas reserves in tight rocks exceed all normally recoverable gas reserves. The quantity of gas per rock unit is at present too small for economic production, and drilling does not bring this gas to the surface. Probably it will be necessary to fracture the rocks. Nuclear explosions which have been used in fracturing tests do not seem to have produced economically satisfactory results until now. More R & D, including the application of other fracturing techniques might be necessary.

D. SHALE

It has been estimated that the largest fossil fuel resource after coal is not oil or gas, but oil shale. Shale rocks often contain some kerogen(1) in solid form. The United States alone has at least 1,800 billion barrels(2) equivalent of shale oil of which 600 billion barrels are considered to be higher grade. This exceeds the conventional oil resources. Extensive oil shale deposits are known to exist in many other countries. Systematic prospection for oil shale has barely started but it is possible that the geographical distribution of oil shale is more equal than that of

1) Kerogen is a complex organic matter which by pyrolysis, is converted into a liquid similar to crude oil. For practical purposes, Kerogen will be called "oil" in this Chapter.

2) 250 x 10^9 toe = 1,100 x 10^{19} joule.

proven oil reserves. Searching more extensively for shale, especially for deeper than surface layers should be a priority task.

Converting shale to liquid oil requires less science than other conversions. It is a process which has been known for more than a century. It is sufficient to heat crushed shale to approximately 450°C in order to liquefy the oil and make it flow out.

What has prevented the large scale exploitation of oil shales are the engineering problems and the costs of moving and heating huge quantities of shale as well as secondary, mainly environmental problems. Techniques which have been developed during the past few years involve extracting the shale and moving it into retorts where the combustion takes place. These techniques require considerable quantities of water and leave a greater volume of spent shale above ground than has been extracted, because the crushing and heating process expands the shale. Approximately three barrels of water are needed for one barrel of shale oil. One and a half of these three barrels are used to consolidate the spent shale, an amount which is expected to be significantly reduced as the spent shale evacuation technology improves.

Spent shale can change the landscape and has other negative side-effects. Its disposal, which may be necessary for environmental reasons, increases the original costs incurred for mining and for building the retorts. Nevertheless, above-ground retorting techniques, including environmental protection costs, became economic in comparison to oil prices at the beginning of 1974.

More recently, a different concept of exploiting shale has been proposed: in-situ or underground oil extraction. Cavities are mined inside the shale layers by traditional mining techniques; in these cavities, shale is crushed by explosives and afterwards heated. The oil gathers at the bottom of the cavities and is pumped above ground. A first pilot test at the end of 1973 was successful.

The advantages of an in-situ process, if it works, are very great. In-situ extraction would seem to be profitable even with low grade shales(1) which form the bulk of all shale deposits but have been considered as too poor for above-ground retorting methods. Unforeseen problems can reduce the flow of oil from one cavity or another but are most unlikely to jeopardise the entire process. In-situ extraction is said to require no water, at least not directly for the extraction process, creates fewer environmental problems, and costs less because it demands less mining and no above-ground retorts. It is true that the oil recovery factor might often be quite low, but low grade shales which would not have been exploited

1) 6 per cent to 8 per cent, or 20 to 22 gallons oil per ton of shale.

otherwise could be profitable even at very low recovery rates if the extraction technology is sufficiently cheap.

R & D on above-ground retorting should continue. It will probably be more widely applied than other methods, at least initially and over the next few years. Above-ground retorting could become the best technology in places where underground water makes in-situ combustion difficult, or for rich surface shales where the possibility of recovering 100 per cent of the oil could make above-ground retorting cheaper per recovered unit. But much emphasis should be paid to the further development of in-situ extraction technologies. If the first positive test is confirmed by others, it could signify a technological and economic turning point in the utilisation of oil shales, with more immediate consequences than coal conversion technologies, at least for countries which have shales.

E. TAR SANDS

Tar sand or oil sand deposits are found in many parts of the world: Canada, the United States and Venezuela, to name a few. Again, global resource data are limited, but known tar sand reserves seem to be less extensive and widespread than oil shales. To recover oil from tar sands, the latter are mined, normally after removing the overburden through strip-mining. Afterwards, the tar or bitumen is separated from the sand, cleaned and upgraded to synthetic crude oil. Several environmental problems can arise during this process. In Canada, development has been slow because of these technical problems and the relatively large capital outlay required. Mining, which is the only technology available in the short term, could recover barely 5 per cent of Canada's large tar sand oil reserves. The technology of extracting oil from the remaining reserves does not yet exist, although R & D on various in-situ recovery methods, including combustion followed by water flooding, or steam injection, is underway.

Tar sand oil recovery is today economic even if oil prices were lower than the present ones. Fifty thousand barrels of oil per day(1) had already been produced in 1972, and more could be produced, which makes tar sand oil recovery one of the most advanced conversion or synthetic fuel processes at present. Nevertheless, R & D on tar sand oil recovery, especially on recovery from low-grade tar sands and from tar sands in greater depths, should continue.

---

1) 2.6×10^6 toe per year.

F. COAL

Many countries wish to restore coal as an energy source as quickly as possible to a more dominating role than it has had during the last ten years. It is impossible to put this wish into practice immediately. A big increase in coal production necessitates the development of cheaper and more efficient methods of coal mining. Moreover, to remove the polluting components of coal and convert coal into more versatile liquid or gaseous fuels, substantial R & D efforts are necessary. Even when these R & D projects are based on known principles and processes, three, five or even ten years of effort will be needed to develop these processes to the point of commercial application. For example, such development involves extensive materials research to adapt conventional equipment to handling corrosive fluids and erosive solids at high temperatures and high pressures.

However, even improved processes will often not constitute optimal solutions. If the aim is to treat and use coal in a completely novel way, R & D lead times could easily last from ten to twenty years. Thus, provision for long term research should be included in coal research programmes.

Much of our existing basic knowledge on coal originates from German research from approximately 1910 to 1940. To improve this knowledge, R & D should be pursued at all levels, including that of fundamental research; coal is a complex mineral and its chemistry is still not fully understood. The science and technology of coal has been in a state of disarray, at least compared to oil or uranium. This reflects a long history of numerous uncoordinated R & D efforts, carried out at various times in various countries, sometimes in emergency conditions, sometimes with little support, and the reduction of R & D efforts during the last decade when interest in coal diminished as a result of low oil prices.

As a result, there are today literally dozens of coal treatment and conversion proposals of various degrees of scientific and technological sophistication; some are ideas only, others have been tested for some parts of the process, but not for others; some proposals have led to pilot plants, but no technology has yet produced, at least recently, commercial quantities of converted coal at acceptable prices, and few have produced processes which can be called mature. To complicate the problem, there are different grades and types of coal, and many technologies apply to some types, but not to others.

If it were possible to foresee which pollution control and conversion technologies would finally be the cheapest, simplest and most efficient, all R & D efforts could be concentrated on them, thus saving money and years of work. At the present early stage, the risk of selecting relatively poor technologies is still high. There is no other way than to follow many parallel paths of R & D, and to slowly eliminate the least successful ones. A number of scientific and technical problems are relevant to different coal technologies and sometimes to other energy technologies as well. R & D on such problems will be useful to whatever technology proves best. Materials research and research on catalysts will be necessary for many coal-conversion, electricity production and other energy technologies.

Coal will ultimately reconquer a much more important position, certainly in countries which have adequate supplies, perhaps also in other countries through increased coal trade. How strong this position will be will depend upon the relative price and availability of coal, marginally and in the short run upon new oil finds, in a few countries upon oil shales, and in the long run upon the speed and scope of nuclear energy developments.

1. Coal Mining

During the last few years, health and safety standards in coal mining have improved in many countries. In the United States, productivity per underground miner is today lower than it was several years ago because coal mining technologies did not improve at the same rate. In other countries, mining technologies improved more quickly.

Since the development of new coal utilisation technologies can only be justified if adequate supply is assured, coal mining R & D should receive priority support. For some years, mining R & D could have a faster and more far-reaching effect than most other coal R & D projects. Coal mining R & D to promote mechanisation and automation is vital for the medium and long term future, as it might be increasingly difficult to find underground coal miners in sufficient numbers. Many scientific and technological advances which have accumulated during the last few years can probably, to some degree, be transferred to coal mining. For example, the drilling of shafts should be improved; in particular the continuous removal of rock material during the drilling phase has not yet been solved efficiently. Coal seams have to be measured and assessed before mining starts, in order to deal with geological faults and acquifers. Considerable progress is possible here, especially through seismic measuring methods.

In mines which are relatively near to the surface, especially in the United States, room-and-pillar mining is still widespread. This method, which consists of leaving natural pillars of coal to support the roof, can be improved. In deeper mines, the most advanced mechanised mining technique of long-wall faces (coal walls) is essentially a combination of three technologies; powered support, a conveyor, and a coal-cutting or ploughing machine. Powered support is roof-support based on hydraulic pillars which advance mechanically. Coal mining and removal are simultaneous; the mining machine drops the coal on a moving conveyor, of which the most delicate component is usually the chain.

The introduction and the improvement of mechanised long-wall face mining are among the most important tasks of coal R & D. The reliability of each of the three technologies must be increased, but it is even more important to improve their integration into a single efficient system, and to adapt it to different and more difficult geological environments, such as thin seams, inclined faces or steep conditions. Today, mechanised systems allow for effective long-wall face mining during approximately 30 per cent of the time, one or two of the most advanced mines even reach 45 per cent.(1) A target of 60 per cent, probably the absolute maximum, might be reached with a steady R & D effort to pinpoint and eliminate production bottlenecks.

Whether advanced mechanisation should eventually lead to complete automation and remote control, is still open to question. In some concrete cases, automation could be very useful. For example, there are plans for preparing the coal underground - mainly washing and crushing it - through an automatic system instead of transporting it to the surface as it is. A more futuristic proposal is that of replacing coal miners completely by mining robots. Technologically, a high degree of automation is probably feasible, but it is not sure that this will be economic in the foreseeable future.

R & D can improve the logistics and technology of underground mining in many other respects. For example, the reliability, size and speed of transport wagons could be increased considerably. Most important are further improvements in working, health and safety conditions. Dust, which can lead to silicosis, explosions and fire provoked by methane, gas escapes and carbon monoxide are still the main dangers, although they claim much fewer lives than a few decades ago. However, safety precautions, early warning systems and techniques to fight fire and other accidents must be improved.

---

1) Eighteen hours of work (three shifts of six hours each) and six hours of maintenance and repair in 24 hours = 75 per cent. During working hours, the machines operate for approximately 60 per cent, and are blocked for the rest of the time because of technical or geological problems. 60 per cent x 75 per cent = 45 per cent.

Better mine ventilation and air-conditioning are desirable. At present, the maximum coal mining depth is 1,200 m., but there are huge resources in greater depth. Only much cheaper and better ventilation and support systems would make it possible to reach them.

Technically less complex than underground mining is surface - or strip-mining of coal. This is done by huge earth-moving machines ("draglines"). Although draglines can still be improved, the main problem for the time being is a capacity bottleneck which has prevented dragline production from keeping up with increasing demand.

As far as surface coal is concerned, the main R & D thrust should still be devoted to the development of reclamation methods for strip-mined land. First experiences in the reclamation and afforestation of strip-mined land have been quite successful. The environment costs of surface coal mining might eventually be smaller than was feared at first.

2. Coal as Boiler Fuel for Electricity Generation(1)

In countries which have sufficient resources, coal will continue to, or will again, and increasingly, be burned in power plants to generate electricity, but most experts agree that the traditional air polluting smoke-stacks should not be allowed to return. In many countries, regulations exist today which lay down a certain minimum height for smoke-stacks, in order to ensure the best possible evacuation of smoke and gas. However, this should not be a permanent solution. For the next few years, R & D to develop clean coal burning technologies will yield quicker results than research on any other coal problem, with the exception of mining. Thus, clean coal burning R & D should certainly receive massive support.

Sulphur oxides, ash and particulates are the three main polluting components of coal which are potentially dangerous to the environment and to public health.(2) Clarifying the climatic and physiological effects of air pollution through coal and other energy sources, is one of the most urgent energy R & D tasks. As these pollution problems fall between various scientific disciplines, they have never been sufficiently investigated, and there is no

---

1) Some details of this and the following sub-chapters are based on: Joint Ad-hoc Working Party of the Energy Committee and the Committee for Scientific and Technological Policy on the Re-assessment of the Role of Coal, OECD, Paris, 19th November, 1974 (mimeographed).

2) Nitrogen oxide pollution which is not specific to coal, but related to all high temperature combustion in air, is not dealt with here.

agreement on scientifically and technologically reasonable or optimal pollution standards. It is not exactly known which of the three polluting coal components - alone or in combination - endangers health, nor are their atmospheric migrations and climatic consequences understood.

In any case, the removal of polluting components is a priority aim of coal R & D. Of the different pollution abatement technologies, none has yet clearly emerged as the best or cheapest. Therefore, research must continue into several directions.

Stack-gas cleaning, Fluidised-Bed Combustion, burning of Solvent Refined Coal or of Low-BTU Gas are some of the methods developed to suppress pollution and to render coal fired power plants more efficient.

Apparently, the most obvious solution is to burn coal conventionally, but to eliminate sulphur, ash and particulates from the stack-gas before it is released into the atmosphere. Technically this method seems to be more mature than any other. Between 1975 and 1976, new stack-gas cleaning methods could become commercially available, according to experts in several countries. Unlike methods used in the past, they will yield an easily disposable, pure sulphur. Eighty per cent of the sulphur of high-sulphur (three per cent) coals can be removed without increasing electricity costs by more than 20 per cent. Many experts consider the figure of 80 per cent sufficient, at least for the time being. Removing almost 100 per cent of all pollutants would be technologically very difficult and extremely expensive.

Still, in the longer run, there might be different and better methods than stack-gas cleaning. Attempts could be made to remove pollutants during, instead of after combustion. Fluidised-Bed Combustion of coal is a boiler-firing technique which removes air-pollutants. It also has other advantages, for example it allows of reducing the size of boilers. A disadvantage of fluidised-bed combustion is, for the time being, linked to the problems of handling solid wastes. At least six different new fluidised-bed combustion concepts are being studied or tested. Perhaps, the prototype of one concept will work at the end of 1974, which could lead to a first full-scale utility boiler late in 1978. Other concepts might be operational techniques in the early 1980s.

Finally, instead of being removed after or during combustion, pollutants can be eliminated even before combustion, by converting natural coal into Solvent Refined Coal (SRC), a clean low-ash and low-sulphur fuel with a melting point of 150-200°C. Cleaning increases the heating value of SRC by approximately 20 per cent compared to natural coal. SRC might replace oil in thermal power stations with little equipment modifications, according to several experts. However, although most remaining scientific and technological problems are likely to be solved within a few years, the costs of SRC are not yet clear, and mass-production is not likely to start before 1980. There are also a number of other, not yet developed concepts for pre-combustion cleaning of high-sulphur coal. Apart from SRC, Low-BTU Gas is for the time being considered as the only other coal-based fuel for electricity generation which could be both clean and economic.

Combustion research is not limited to the cleaning of coal or stack-gas. For example, research on preparation and combustion of pulverised coal and flame research are important contributions to the improvement of coal combustion technologies.

3. Low-BTU Gas

After the mining and cleaning of coal, coal gasification should become a third R & D priority. The long term future could belong to versatile coal-fuels at least as much as to solid coal. Many, though not all, experts believe that synthetic gas will become the most important coal-based fuel, and that therefore, most emphasis should be laid on the development of gasification technologies. There are several reasons for this. Gas is often easier to use than liquid or solid fuels. Pipeline transport of natural gas is a cheap energy transport, whereas railway transport of coal is the most expensive one with electricity transmission costs ranging between the two. In addition, gasification might be simpler and, according to many experts, cheaper than liquefaction technologies, simply because gasification is chemically speaking a more radical process than liquefaction.(1) Ideally, a conversion process should produce a clean fuel of high calorific value, comparable to natural gas. However, coal conversion creates complex chemical, physical, metallurgical and engineering problems, which explains why there are so many R & D programmes under way, without a clear commercial solution in sight for the immediate future.

Low-BTU Gas for direct use (150-200 BTU per cubic foot)(2) has become an R & D target because it will be cheaper to produce than

---

1) In simplified form: in gasification, coal, steam and air are first broken down into their smallest components, and subsequently reassembled into a new molecule. In liquefaction, molecules are only partly broken, and hydrogen has to be added. Chemically less radical modifications can be technically more difficult and expensive.

2) $1.6 \times 10^5 - 2.1 \times 10^5$ joule per ft^3; or 1,300 - 1,800 Kcal per $m^3 = 5.6 \times 10^5 - 7.5 \times 10^5$ joule per m^3.

high-BTU gas. The chief use foreseen for it is as clean fuel in combined gas-turbine-steam-turbine power plants, or "advanced power cycles".(1) It is relatively uneconomic to transport low-BTU gas, because of its high volume-to-heat ratio. Therefore, power plants will have to be built in combination with, or in proximity to gasification plants. Thus, low-BTU gas will help to open the way to new gas turbine technologies.

Within these geographical and technical limitations, low-BTU gas could become a dominant electricity generation fuel. Certainly, there is no technical limitation to burning more expensive high-BTU gas or liquefied coal under boilers. However, as long as conversion efficiencies from fuel to electricity remain as low as they are today only the cheaper coal-based fuels will be competitive.

A small number of low-BTU gas conversion processes are presently being studied. One of them is an in-situ or underground gasification process which would make coal-mining and transport superfluous, at least in surface-coal areas. The still unsolved and perhaps unsolvable problem is to control, to move and to collect the gas which develops through in-situ combustion. A commercial low-BTU gas for power stations could be available by about 1980.

Apart from this low-BTU gas, research is going on to produce a "synthesis gas".(2) In the United States, and perhaps only there, synthesis gas might be used directly as boiler fuel for electricity generation. Its main, and in some countries unique, use will be intermediary; it will be the basis for the production or "synthesis" of high-BTU gas. Several different conversion processes into synthesis gas are being studied. As they received attention much later than other coal conversion technologies, they are still in an early development stage, and commercial products are not yet in sight.

1) To increase the efficiency of coal-based power plants, fuel improvements are obviously not the only solution. One can also try to increase conversion efficiencies, through combined power cycles, or through MHD, or one can improve the equipment, for example the boiler design. Cf. Chapter IV.

2) In American and British terminology, all coal-gas that is not methane or equivalent to methane in heating value is called "low-BTU gas", including this synthesis gas. German terminology distinguishes between:

   a) Low-BTU Gas, approx. 1,500 kcal per m^3 (63×10^5 joule/m^3);

   b) Synthesis Gas, approx. 2,500 kcal per m^3 (100×10^5 j/m^3);

   c) Town Gas, approx. 4,200 kcal per m^3, traditional side-product of coal coking, (180×10^5 joule/m^3);

   d) Natural Gas, Substitute Natural Gas (SNG), "High-BTU Gas", approx. 9,000 kcal per m^3, (380×10^5 joule/m^3).

4. High-BTU Gas

There is one process by which high-BTU, or "Pipeline Quality Gas", or SNG, could probably soon be produced in commercial quantities. The old Lurgi gasification process, technically improved, converts coal into a 350 BTU per cubic foot gas (3,100 kcal/m^3) which can be enriched or methanated to 500 (4,550 kcal/m^3) and then to 900 BTU per cubic foot (8,000 kcal/m^3)(1). Several big Lurgi plants are under construction. They are expected to deliver in a few years larger quantities of synthetic gas. However, the present Lurgi process is not the best, nor the ultimate gasification technology. R & D must be applied to improve it and to find cheaper methods of making synthetic gas of the same heating value as natural gas, which is 900 to 1,000 BTU per cubic foot (8,000 à 8,900 kcal/m^3).(1)

High-BTU conversion mainly consists of three steps: gasification, purification to remove pollutants, and methanation to upgrade the heating value. Gasification and purification are still expensive steps. Gasification requires much heat (more than 900°C for bituminous coal); in gasification processes now used, a third of the coal must be burned to provide the heat for the gasification of the remaining two thirds. High temperature gas cooled nuclear reactors (HTGR) which are being developed now could provide this heat. Such a technological breakthrough would reduce coal conversion costs by 20 to 30 per cent and increase the coal reserves available for conversion by a third. A combined high temperature reactor - coal conversion plant is a very appealing solution which would deserve more interest and R & D support. At the present rate of progress, a demonstration plant will be in operation in the late 1970s, a full-scale plant in 1982-83 and commercial plants approximately by 1988.

---

1) a) 350 BTU/ft^3 = 3.7 x 10^5 joule/ft^3, or
= 3,100 Kcal/m^3 = 130 x 10^5 joule/m^3;

b) 500 BTU/ft^3 = 5.3 x 10^5 joule/ft^3, or
= 4,550 Kcal/m^3 = 190 x 10^5 joule/m^3;

c) 900 BTU/ft^3 = 9.5 x 10^5 joule/ft^3, or
= 8,000 Kcal/m^3 = 330 x 10^5 joule/m^3;

d) 1,000 BTU/ft^3 = 10 x 10^5 joule/ft^3, or
= 8,900 Kcal/m^3 = 370 x 10^5 joule/m^3.

After gasification and purification, it is the methanation step which might need some additional research, although methanation has recently been demonstrated on a substantial scale.

Approximately ten different, new high-BTU gasification processes are presently being investigated: four are being or will soon be tested on a larger pilot-scale. Commercial plants based on new post-Lurgi technologies, but not HTGRs, might be ready in 1981-1982.

5. Liquid Fuels from Coal

Coal liquefaction is no novelty. Coal can be converted into heavy or light liquids, including gasoline, according to process and to the quantity of hydrogen added. Germany during the war liquefied coal in large quantities, but the techniques of that time are generally regarded today as too inefficient and expensive(1). South Africa is at present the only country which produces a synthetic liquid in commercial quantities.

All coal liquefaction goes back to one of four basic technologies which have all been known for decades; tar production through traditional coal-coking or pyrolysis, solvent refined coal extraction (Pott-Broche process), coal hydrogenation (Bergius-Pier process) and gasoline synthesis (Fischer-Tropsch process).

New liquefaction R & D consists of modernising one or several of these well-known processes. New catalysts can be developed, production engineering of the 1930s and 1940s can be upgraded to the present state of the art, and new methods to produce cheaper hydrogen, which would reduce liquefaction costs considerably, can be investigated. One of the main unsolved problems is the limited applicability of liquefaction techniques. They can process certain types of coal, but not others.

Nevertheless, the uncertainties of coal liquefaction are due less to scientific and technological obstacles, than to uncertainties related to the future end use of synthetic oils. Converting coal into heavy heating oil, mainly SRC, for electricity generation, or into light heating oil for space heating, will probably make sense and might be economic. Whether the same can be said of a possible, but more expensive, synthetic gasoline production, is not at all certain. If coal has to be liquefied into a transportation fuel, the cheapest and best solution of the future might well be to

1) American sources have mentioned $12 per barrel as a probable price for liquefied coal, if an obsolete Fischer-Tropsch process had to be used today.

apply high temperature gas cooled reactors to convert coal, not into oil, but into methanol.(1) Methanol can replace gasoline with minor engine modifications.

Today, at least four improved liquefaction processes for heating oil are in the development stage. Commercial production could start by the early 1980s. Oil companies have indicated that they are working on new liquefaction techniques which they hope to make competitive, but no details are available.

6. Other Coal R & D Problems

There is scope for many other R & D programmes in the coal sector. Transport of coal and coal products can still be improved by additional R & D. Considerable progress has recently been made in transporting solid coal by pipelines, with the help of water or oil ("coal-water" or "coal-oil-slurries"). Additional progress is possible in this and other technologies and might even be necessary to permit increased coal trade between countries, or transport inside large countries such as the United States or Canada. A problem which preoccupies industries and countries which have little or no coal is the location of future coal conversion plants. Political criteria will certainly influence decisions, but so will the comparative costs of transport.

Coal has to be prepared before it can be used; it must be "washed" to remove stones, and it must be crushed before gasification or liquefaction. More coal preparation R & D is necessary.

Many coal conversion technologies produce by-products, apart from the gaseous or liquid fuels which are their main aim. An efficient, economic utilisation or disposal of these by-products could in some cases make the difference between a competitive and non-competitive conversion technology. This problem deserves more R & D. In particular, new processes can reduce the polluting effects of the coking of coal for steel production and make it more efficient and economic. R & D on this has been, and will continue to be profitable.

Finally, until today, the best known coals have been the ones with the highest heating value. New R & D on many lower quality coals which can be found, and corresponding R & D on equipment modifications which might be necessary for the use of these coals, will eventually be very profitable. For example, countries such as

1) At 950°C from a HTGR, one ton of coal plus water steam produces approximately two tons of methanol, equivalent in heating value to one ton of gasoline. Thirty-five per cent of the HTGR heat is chemically bound in the methanol. The carbon conversion yield is very high.

Finland, Ireland and Sweden have large peat reserves and have successfully burnt peat for electricity generation, or are investigating new peat burning technologies.

G. ECONOMIC PROSPECTS OF NEW OIL AND SYNTHETIC FUEL SOURCES

The economic value of new technologies in the fossil fuel sectors will be measured by the speed with which they will relieve energy shortages, and by the quantity and price of the new energy resources which they will be able to provide in the long run. Indications on what might be possible or probable can be given today, but no more.

Off-shore oil - New fields and off-shore fields which were known at the end of 1973 will provide a considerable proportion - optimistic experts say 50 per cent - of all oil for OECD countries by 1980. The Alaskan and North Sea fields alone were, according to 1973 forecasts, expected to produce at least five million barrels per day by 1980, and this figure could well increase. Other, less publicised fields which have been discovered might add to this figure. Increased drilling is likely to reveal new off-shore fields from 1974 on. New oil which has not been discovered in 1974 will not make a very big contribution to supplies before 1980. But after 1980, yet undiscovered off-shore oil is expected to increase supplies very considerably, even if present hopes, based on geological structures, come only partly true. Production costs(1) of off-shore oil vary. Industry estimates that landed costs for both North-sea and Prudhoe Bay (Alaska) crudes will range between $1.25 and $1.50 per barrel by 1980, at least for fields which are already under development.(2) Costs for oil fields in deeper waters are likely to be higher; for a water-depth of approximately 300-400 m., costs are likely to approach $4 per barrel. Deep-sea oil, produced at a water depth from 300-400 m. to 2,000 m., could cost between $5 and $7 or $8. Estimates are more difficult in this case, as drilling and production technologies for this depth are mostly not yet available. On the other hand, improvements in recently developed prospection, drilling and production technologies, and mass-production, could bring unit costs down during the 1980s.(3)

---

1) Costs include technical unit costs and amortization of capital costs. All dollar costs in this chapter are in 1973 dollars.

2) Cf. Report on Conventional Crude Oil Reserve Position and Potential for an Increase in Oil Production in OECD Member Countries by 1980 and 1985, OECD, Paris, 18th April, 1974 (mimeographed).

3) 5 million barrels per day = 260 x 10^6 tep per year;
$1.25 - $1.50 per barrel = $2.1 - $2.5 per 10^{10} joule;
$4,5,7,8 per barrel = $6.7, 8.4, 11.8, 13.5 per 10^{10} joule.

Secondary and Tertiary Recovery - Improvements in secondary and tertiary recovery will affect supplies, although two or three years may elapse from the application of more sophisticated recovery methods to the time when additional oil appears; this makes sense mainly after the normally recoverable oil has been produced, which for some oil fields, will be in decades only. Another result of recovery improvements will be to stretch out oil supplies for many more years than had been foreseen originally. For the time being, old fields which have been abandoned are starting to produce again. It is difficult to foresee how much oil they will produce by 1980, but the quantities could in some cases be substantial. Between 1974 and 1980 fields which have been abandoned after they had yielded 30 per cent of their oil, might produce another 7 per cent, or almost a quarter of all oil produced since the very beginning. Additional costs per barrel for secondary and tertiary recovery are expected to range between $0.25 and $1.50, according to recovery method and to the proportion of oil that has already been taken out. Hence, total costs per additional barrel can be anything between less than $1 and $5, according to field.(1)

Shale - Calculations have until recently been based mainly on above-ground retorting technologies. Forecasts on quantities have been vague; figures have varied between 100,000 and more than one million barrels per day in 1985. Cost estimates for above-ground technologies vary between $4 and $6, even $7 per barrel. The difference is due to different estimates of capital costs, and of environmental costs. If in-situ extraction technologies work as well as expected, which should be better known by the end of 1974, these figures will change. As capital investments for in-situ recovery are smaller, some extraction could start soon - say in three years. Cost estimates vary between approximately $3 and $5 per barrel.(2)

Tar sands - The role that tar sands will play in oil production from 1980 on depends, especially in Canada, upon the rate at which capital, labour and other supplies can be marshalled. Owing to the oil shortage, the Canadian economy might have no alternative but to pay the price necessary to make the production of petroleum from tar

---

1) $0.25 - $1.50 per barrel = $0.4 - $2.5 per 10^{10} joule;
   $1 - $5 per barrel = $1.7 - $8.4 per 10^{10} joule.

2) 100,000 - 1 million barrels = 50 x 10^{5} - 50 x 10^{6} toe per year.
   $3 per barrel = $5.1 per 10^{10} joule;
   $6 per barrel = $10.1 per 10^{10} joule;
   $4,5,7 per barrel, cf. Footnote 3, page 45.

sands profitable. A cost of approximately $4 per barrel(1) has been mentioned as a possibility.

Fuels from coal - Estimates on the future quantity, price and development time of different synthetic fuels from coal have been varying between and within countries, and they are still changing as accelerated R & D efforts reveal that the problems which have to be solved can be bigger or smaller than was expected before 1974. Beyond the technical, there are several economic and political uncertainties which render forecasts difficult. The future price of coal, and the financing mechanism for future coal conversion plants, are among those uncertainties. Coal prices, which vary considerably between countries, will depend upon the international coal trade, national subsidies and many other factors, apart from mining technologies. Moreover, synthetic fuel prices can vary by at least 10 or 20 per cent according to whether plants are built with a utility type financing, or on a 100 per cent equity capital basis. Nevertheless, it is possible to estimate to some degree the order of magnitude of future synthetic fuel prices,(2) as well as probable or approximate development times.

i) Clean, coal-based fuels for Electricity Generation, excluding HTGR Technology

a) Normal coal and subsequent stack-gas cleaning: commercial in one to two years (1975-76);
b) Solvent Refined Coal (SRC): commercial in approximately four to six years (1978-81);
c) Fluidised-Bed Combustion: commercial in approximately four to seven years, according to process (1978-82);
d) Low-BTU gas with steam and gas turbines: commercial in approximately seven to twelve years (1980s).

If calculated on the basis of United States coal and world oil prices of 1974, each of the four systems could, if available, already be competitive with petroleum-based electricity generation. New stack-gas cleaning methods could increase electricity costs by approximately 20 per cent, at least at the beginning, compared to electricity costs from traditional coal power plants. Solvent refined coal, produced from United States coal, could cost between $5 and $7 per barrel-equivalent.(3) At the beginning and based on first generation equipment, it is expected to cost more, but mass

1) = $6.7 per 10^{10} joule.

2) The following figures are largely based on the "Reassessment of the Role of Coal", OECD, 23rd April, 1974, document cited above, page 38. Conversions into barrels are calculated on the basis of one barrel crude = 5.8 million BTU or 1,461,600 Kcal.

3) $8.4 - $11.8 per 10^{10} joule.

production and technological improvements could bring costs down later on. SRC can probably replace fuel-oil for existing thermal power plants with little equipment changes.

In the medium term, Low-BTU gas could be one of the cheapest electricity sources, if the development of the gas turbine and steam turbine technology does not run into unforeseen troubles. Low-BTU gas from United States coal is expected to cost between approximately $3.20 and $4.60 per barrel-equivalent.(1) If world oil prices have not come down substantially by the time these technologies are commercial, oil will not be competitive any longer as a power plant fuel.

ii) High-BTU Gas

High-BTU gas from coal as replacement of natural gas will be produced in larger quantities from approximately 1978 on, on the basis of Lurgi technologies. The costs of this synthetic gas in the United States will be approximately $8 per barrel-equivalent. Improved gasification technologies could lead to new plants from 1981-1982 on. Future costs in the United States are estimated at between over $6 and $9 per barrel-equivalent, at least for the beginning(2).

iii) HTGR - Technology

The use of cheap process heat from High temperature gas cooled reactors for coal conversion could mean an economic breakthrough. It could reduce the costs of Low-BTU gas, High-BTU gas, SRC and light liquids from coal by an additional 20 to 30 per cent compared to the preceding figures. Such a breakthrough is now expected for the 1980s. As transportation fuel, Methanol could profit most from this. Available technologies convert methanol from United States coal for $10 to $12 per barrel-equivalent; HTGRs could reduce costs to approximately $6 per barrel-equivalent.(2)

Quantitative Prospects

It is difficult to estimate the quantities of new oil and of synthetic coal fuels which will be available in the future. However,

1) $5.4 - $7.7 per 10^{10} joule.

2) $4 per barrel = $6.7 per 10^{10} joule ;
   $5 " " = $8.4 " " " ;
   $6 " " = $10.1 " " " ;
   $7 " " = $11.8 " " " ;
   $8 " " = $13.5 " " " ;
   $9 " " = $15.2 " " " ;
   $10 " " = $16.4 " " " ;
   $12 " " = $20.2 " " " .

it is certain that already for $4 to $5 per barrel(1), new technologies can make large quantities of new oil available from off-shore fields, secondary or tertiary recovery fields, shales or tar sands. By 1980, these different - often indigenous - sources together will provide OECD countries with several million barrels per day, and new supplies might increase steeply during the 1980s.

The supply of gaseous and liquid fuels from coal will probably not reach similar quantities before 1980. Synthetic coal fuels might only in the early 1980s add up to an equivalent of one million barrels per day. Apart from technology and government policies, much will depend upon future oil prices. For $5 to $7 per barrel oil,(1) coal conversion technologies could, probably not at the beginning but in the long run, become economically attractive, certainly with United States coal prices. For $7 per barrel(1) many known and projected conversion technologies are competitive and could produce quickly increasing quantities of synthetic fuels during the late 1980s, again based on United States coal prices.

The main difficulty in forecasting total supplies from new oil and synthetic fuel sources stems from the inter-dependence of most of these sources. Many will be produced by the same oil and energy companies, in factories which will make considerable demands on the same construction and engineering companies. Therefore, supply and cost of capital and skilled manpower, and the production capacities of construction companies, might influence fuel prices more than R & D, at least for a certain time. Of course, governments can modify all these factors through direct support, taxation or other measures. Nevertheless, the simultaneous demand of many companies for the same resources might lead to unexpected bottlenecks, delays and higher construction costs in the energy sector, at least for a few years. Afterwards, in the 1980s, economies of scale could bring costs down again.

In the past, market economies have reacted to a sudden price increase or a lasting shortage of an important commodity by increased production efforts and by technological substitution of that commodity. This reaction has, more often than not, turned the shortage, after a number of years, into a surplus, bringing about a price collapse. Even if historical experiences have only limited value, new oil finds, oil production by new technologies, the development of synthetic fuels, and energy conservation measures could well change present trends again, maybe to the point where international oil prices will have to be very substantially reduced compared to their 1974 level. Of course, one cannot forecast the development of the many political and economic factors which

1) See footnote (2) on page 48.

influence demand, supply and price of oil, sometimes even more than technology. What one can forecast, is that for oil prices between $4 and $6 or even $7 per barrel(1) technology could open up new oil and other fossil fuel sources which are big enough to greatly modify the entire energy situation and which were financially and technologically out of range before 1973.

Finally, one of the most vital general contributions which the R & D system has to make to solve energy shortages, is the development of better mass production and standardization techniques, and the training of many more prospection, mining and chemical engineers. Standardization in fuel, for example in coal conversion factories, and the supply of a sufficient number of engineers can make all the difference between early or late, cheap or expensive replacement fuels, at least during the 1970s and early 1980s.

1) See footnote (2) on page 48.

II NUCLEAR ENERGY

As a result of a vast R & D effort which has gone on continuously for more than 25 years, nuclear energy is now the energy source best able to take over from hydrocarbon fuels in the large-scale generation of electricity and give satisfaction from the economic, environmental and safety standpoints. While hydrocarbons must be considered also as extremely important direct sources of materials and possibly of foodstuffs, uranium and thorium can only be used at present for producing energy.

However, so that the long-term role of nuclear energy can be determined and nuclear strategy worked out for the next decades, it is essential and urgent to add substantially to our present inadequate knowledge of the world's uranium and thorium resources. It is particularly important that we should have more accurate data about available resources in terms of recovery costs, without neglecting those of which the cost is too high for them to be recovered in the near future. It must also be borne in mind that intensified prospecting for uranium is already necessary, even in the relatively short term, in order to ensure continuity of supplies from 1980 onwards.(1) A prerequisite for the accelerated development of nuclear energy now planned in many Member countries is a substantial increase in R & D covering all types of uranium and thorium resources and the methods of their recovery. The aim of this R & D effort should not only be to discover new high-grade deposits but also enable uranium to be extracted from sources such as phosphates, granitic rocks or even sea water.

A. NUCLEAR REACTORS AND FUELS

1. The present generation

New nuclear electricity generating capacity to be introduced in the next ten years will mainly involve reactor types which already exist: principally light water reactors (LWR) (these will account for 90 per cent of installed nuclear capacity in 1980) and, to a

1) Cf. Uranium-Resources, Production and Demand. OECD, Paris, 1973.

much lesser extent, heavy water reactors (HWR). Although these technologies are already proven, it should be possible by dint of continued R & D to improve them still further, especially as regards their safety and environmental aspects and the reliability of the non-nuclear components. An attempt should also be made to introduce a greater degree of standardization in order to facilitate construction and simplify licensing procedures, although standardization is of course limited by the conditions specific to each nuclear power plant.

However, while these operations are primarily industrial ones, R & D should be intensified mainly as regards those parts of the fuel cycle outside the reactor - prospecting, production and conversion of uranium, enrichment, fuel fabrication, reprocessing of irradiated fuels, transport and processing of wastes. Existing plants will rapidly become inadequate to handle the increasing amounts of nuclear materials which will result from nuclear energy development, as planned in most Member countries; it is therefore necessary to take a fresh look at the implications of such development for the whole fuel cycle, not only with regard to the industrial capacity which will have to be installed but also from the environmental and safety standpoints.

In particular, the expected rise in the number of light water reactors also calls for a considerable increase in uranium enrichment capacity.(1) This is especially important because, even on the basis of the most optimistic assumption for the rate of introduction of fast breeder reactors, the effects on the demand for enriched uranium would hardly be felt before 1995. At present, there are two main enrichment techniques which should be considered as complementary rather than competitive. Of the two, only the gaseous diffusion process has been proved on an industrial scale. The more recent technology of the ultra-centrifuge has two main advantages: first it requires appreciably less power, and secondly, it permits a much more gradual increase in enrichment capacity as demand develops. However, solutions will have to be found to the problems of producing centrifuges in large numbers, expecially those components which require very high precision engineering. Apart from these two technologies, longer term research should be encouraged into other possible methods of isotopic separation, in particular separation by laser, which could possibly by introduced towards 1990.

1) Of course, this remark does not apply to natural uranium reactors such as the CANDU reactor which has been developed in Canada. The introduction of this latter type of reactor requires the production of heavy water but the technologies for heavy water production are relatively less complex and costly than the uranium enrichment technologies.

There is also a good case for rapidly expanding R & D on plutonium recycling(1); according to certain estimates, this procedure could increase the efficiency of the use of natural uranium from 0.5 per cent to about 0.7 per cent(2) and at the same time appreciably reduce the need for more enrichment capacity.

2. Advanced thermal reactors

The concern for making more effective use of uranium resources and for being able to use thorium as well is the main reason for research now going on in many countries into various types of reactor, some of which could be introduced as from 1980. One type, the Advanced Thermal Reactor, is now under study in Japan; it will be moderated by heavy water, cooled by boiling light water and as fuel would use either 1.5 per cent enriched uranium or a mixture of plutonium and natural uranium. Another example is suggested by the current research in Canada into the use of the thorium-uranium 233 cycle in heavy-water reactors. The possibility of using thorium as fertile material is also one of the advantages of the Self-Sustaining Light Water Reactor (also called the Light Water Breeder - LWBR); at start-up, this reactor would require more enriched uranium than light-water reactors, but during its lifetime of about 40 years, it would burn some four times less uranium oxide and would require nearly three times less separative work.

It must be stressed that the development of all reactor types using thorium as fertile material calls for an appreciable increase in R & D on the whole thorium-uranium 233 cycle. This is particularly important for the rapid development of the high temperature gas-cooled reactor (HTGR) which deserves special attention because of the particularly substantial benefits that may be expected of it. In addition to the advantages of having a higher fuel utilisation and of employing thorium, the safety requirements prescribed for all nuclear reactors seem to be more easily met in the case of HTGR.

Its main interest lies in the fact that the temperature of the helium coolant is appreciably higher than in all other types of reactor. In commercial reactors now on order, the coolant temperature would appear to be of the order of 750°C, a value which would already guarantee a thermodynamic efficiency appreciably higher than in existing nuclear power stations. Later on it should be possible to increase temperatures still further (900 - 1000°C); the AVR experimental reactor at Jülich in Germany (15 MWe) reached a temperature of 950°C in February 1974. This would permit the use of direct cycles

1) See A.5 "General Remarks", p.56

2) Although natural uranium contains 0.7 of the fissionable uranium isotope (U-235) it is not possible for technical reasons to use more than 0.5 per cent without plutonium recycling.

or combined cycles using high-temperature gas turbines. Most important, temperatures of this order hold out the prospect of using the heat directly in certain metal and chemical industries (as envisaged for example, for the gasification of coal(1) and the production of hydrogen), and hence would enable energy resources to be far more efficiently used.

For all these reasons, a marked R & D effort should be devoted to HTGR reactors which might well be expected to develop substantially as from 1980.(2)

3. Fast breeder reactors

The basic objective of the very substantial R & D programmes on fast breeder reactors is to improve the utilisation of nuclear resources by a factor of 40 to 50. A noteworthy fact is that all countries active in this field - France, Germany, Japan, the United Kingdom, the United States and the USSR - have concentrated virtually all their efforts on the same type of fast breeder, the LMFBR, a fast reactor cooled by liquid sodium, using plutonium as the fossil fuel and natural uranium as fertile material. So far the greatest progress has been achieved by the Phenix reactor (250 MWe) at Marcoule, France, which was connected to the national electric power network at the end of 1973 and has been operating at full capacity since March 1974. If development is accelerated, the first power stations on an industrial scale (1,000 MWe and above) could appear on the scene by 1980; their commercial expansion would then be substantial as from 1985.

Although in all countries the greater part of expenditure on fast breeders is earmarked for the LMFBR, some R & D effort should be maintained to keep open and explore further the longer-term options which include the gas-cooled fast breeder reactor (GCFBR)(3) and secondly the molten salt breeder (thermal breeder using thorium as fertile material).

4. Economic prospects(4)

From 1972 onwards, electricity generated by power plants using light-water reactors was competitive with that generated from other sources; this is shown in the table below which gives the lower and

1) Cf. Chapter I, Part II

2) This R & D effort should include research on the availability of the helium resources which would be required for the construction of an important number of HTGR reactors.

3) It may be noted here that techniques necessary for the development of the HTGR would be applicable to this type of breeder.

4) This question is covered in detail in the Nuclear Energy Agency paper, "Nuclear contribution to future energy needs in OECD countries", NE(74)5. All the figures given here are taken from this paper.

upper limits on total costs (in mills/kWh)(1) of electricity generated from various sources for the year 1972 in the OECD area. The recent increases in oil prices have of course confirmed the advantages of nuclear energy.

	Nuclear	Natural Gas	Oil	Coal
Lower Limit	9.2	7.5	9.5	9.5
Upper Limit	9.8	10.3	12.4	15.4

The principal future advantage of nuclear energy lies in the fact that the relative cost of nuclear fuels is much less than that of fossil fuels. Investment in the fuel cycle accounts for only about 10 per cent of the capital cost of the nuclear power station and the natural uranium itself represents only about one fourth of the total cost of the fuel cycle. Thus, even if the cost of natural uranium were later to rise appreciably, the effect on the cost of electricity generated in nuclear plants would be very small.

On the other hand, the capital costs of nuclear power stations have been rising rapidly in recent years, not so much for economic reasons as owing to authorisation procedures, which considerably lengthen the time needed for construction and introduce substantial administrative expenditure. Authorisation procedures vary appreciably from one country to another, thus largely accounting for the substantial differences found in the capital costs of nuclear power stations. Bearing these differences in mind, a rough estimate of the average of the specific capital cost for a nuclear power station might be 300 \$/kWe (in 1972 dollars); if interest charges during construction are added, the total capital cost of a 1,000 MWe plant to come into service in 1980 would be of the order of \$390 million (in 1972 dollars). Taking capital expenditure on the remainder of the fuel cycle to be equivalent to 10 per cent of that for the power plant, the cost of electricity generated by nuclear plant in 1980 (in 1972 dollars) would be between 13.4(2) and 14(3) mills/kWh, where the lower limit is based upon an assumed 50 per cent rise in the current price of uranium and the upper limit on a 100 per cent rise in this price.

The foregoing concerns light water reactors. At present no accurate data are available for the other reactor types likely to be developed as from 1980, particularly high temperature reactors and fast reactors. One of the objectives of R & D on these

1) If expressed in $\$/10^{10}$ joule these costs are -

	Nuclear	Natural Gas	Oil	Coal
Lower Limit	25.5	20.8	26.0	26.3
Upper Limit	27.2	28.6	34.4	42.7

2) 37.2 $\$/10^{10}$ joule.

3) 38.8 $\$/10^{10}$ joule.

reactor types is in fact to reduce their cost, which will depend indirectly on non-technical factors such as safety regulations and authorisation procedures which can lead to technical modifications and additional delays. The most that can be said is that the cost of generating electricity in plants using these reactors will be of the same order of magnitude as it is for light water reactors and that the differences in generating costs between different reactor types will probably be less than the differences in the cost of electricity generated by nuclear plants and that produced in oil or coal-fired stations.

5. General remarks

Quite apart from the R & D programmes related specifically to each type of reactor, of which safety research does of course form an integral part, the overall development of nuclear energy requires a sustained R & D effort on materials. Most important, it is absolutely essential to intensify R & D on all aspects of the various fuel cycles.

Moreover, consideration of the different types of reactor should not be limited to comparing their economic characteristics. Questions of safety, authorisation procedures, prospects for uranium supplies, industrial potential and political considerations are at least as important as strictly economic factors.(1) The different types of reactor should also be considered complementary to a very large extent. This aspect is particularly clear with regard to the fuel cycle; for example the use of thorium may be considered as complementary to that of uranium; another example is the use of plutonium produced in light water reactors as an initial (start-up) fuel for fast breeder reactors. In this context, there is a good case for working out what would be the best 'reactor mix' for nuclear generating plants in each country.

More generally, it is essential for R & D strategies in the nuclear field to be based on a review of all available nuclear systems.

B. THERMONUCLEAR FUSION

Like fast breeder reactors, thermonuclear fusion offers the prospect of utilising an extremely abundant source of energy. It must be stressed however that this is not an infinite source: the only fusion process which today appears feasible is the

1) It is also necessary to mention the problems associated with the siting of nuclear power plants, which should be seen in connection with the more general issue of the siting of power plants (See Chapter IV and VI, Part II).

deuterium - tritium reaction; while deuterium supplies obtainable from seawater may be regarded as virtually inexhaustible, tritium is obtainable from lithium, a relatively less abundant element (0.006 per cent in the earth crust and 0.1 pp.M. in sea water).

It should be pointed out that development of thermonuclear fusion will also lead to safety and environmental problems but although no definitive statement can be made at present on this matter, for lack of information available, these may be less severe than in the case of fission.(1)

It must also be remembered that the scientific feasibility and, to a still greater extent, the engineering feasibility of fusion are both still uncertain,(2) which is not the case with breeder reactors. With a view to achieving the first of these objectives, two principal avenues are now being explored: magnetic containment and fusion by laser.

Magnetic containment has been investigated for more than 20 years and has made remarkable progress as a result of parallel advances in the theoretical physics of plasmas and in the development of increasingly powerful machines, amongst which the Tokamak has given best results. All the same, the rate of progress is limited in particular by the time needed to build these complex machines and to interpret the results on which the subsequent generation of machines must be based. An optimistic view is that the scientific feasibility of fusion with magnetic containment might be demonstrated towards the middle of the next decade.

Laser fusion by compression of a deuterium-tritium pellet by a laser beam or several simultaneous beams, is a much more recent approach. It has made such spectacular progress that many experts believe the scientific feasibility of fusion could well be demonstrated sooner by this method than by magnetic containment, on condition that increasingly powerful lasers are developed fairly quickly. As compared with magnetic containment, laser fusion might have several advantages: in particular, greater flexibility and smaller power plants. Without denying this possibility, it is still necessary to beware of excessive optimism since the theoretical and experimental understanding of the interaction between a laser beam and an extremely dense plasma is still far from complete.(3)

1) A recent study by Dr. W. Seifritz of the Institut Fédéral des recherches en matière de rèacteurs (Switzerland), entitled Stand und Entwicklungstendenzen der Fusionsreaktortechnologie provides some interesting indications on this subject.

2) While fusion must be regarded as one of the unconventional sources of energy covered in the following chapter, it was decided to deal with it here owing to its scientific and institutional links with nuclear energy.

3) In this connection, it may be recalled that in the early days of research on magnetic containment, hopes of rapid success were dashed many times because of the inadequate theoretical understanding of plasma instabilities.

Although the scientific feasibility of fusion produced by either method is still uncertain, many problems of design and technology (in particular of materials) common to the two approaches can be identified and deserve to be investigated, at least on paper, with a view to providing an adequate basis for the research that would be necessary to attain the second objective, the engineering feasibility. In any event, even taking the most optimistic view, it is highly improbable that thermonuclear fusion can be applied to the production of energy before the year 2000.

C. SAFETY, ENVIRONMENT AND PUBLIC OPINION

Among the many problems raised by the development of nuclear energy on an industrial scale, the most important ones are probably those associated with radiation protection, the safety of nuclear installations and the management of radioactive wastes. It should be borne in mind that these questions have always been a component part of R & D dealing with civil applications of nuclear energy, and that on this account the nuclear industry has probably the best record where safety is concerned. Developments now planned for nuclear energy will however call for greater numbers of nuclear installations and for such amounts of nuclear materials that an intensification of research into these problems is absolutely essential.

Research into the effects of radiation should be continued. It would be desirable in particular to carry out epidemiological studies on the effects of low radiation doses.

Safety research should of course continue to be a fundamental part of R & D programmes on all types of reactor and should receive increased support. Moreover, it is equally important that R & D be appreciably intensified on the safety aspects of all other parts of the fuel cycle.

As regards radioactive waste management, the main problem is that of high-activity wastes, i.e. fission products (Sr-90 and Cs-137) but especially Pu-239 and other transuranium elements, the half-lives of which are much longer (24,000 years for Pu-239 compared with about 30 years for Sr-90 and Cs-137). The methods now in use may be regarded as adequate temporary solutions pending the development of satisfactory definitive methods required for the long-term development of nuclear energy. Several options have been envisaged as ultimate solutions such as disposing of wastes in different kinds of deep geological formations. Some others are still at the conceptual stage, namely launching into space, separating out transuranium elements, transmuting transuranium elements by neutron irradiation, and so on. Taking into account the different level of development of the various methods, none of these have been sufficiently investigated yet to go unchallenged. This should therefore be the prime objective of research in this field, while any new possibilities that might come up should also be examined.

It must be added that nuclear safety and radioactive waste raise not only scientific and technological problems but also questions of daily management which are at least equally important. Another serious problem associated with the rapid development of nuclear energy is the question of security, e.g. the risk of sabotage of nuclear plants or diversion of nuclear materials such as plutonium or highly enriched uranium. Although science and technology can help reduce these risks by developing systems which render these illegal actions as difficult as possible, their role is nevertheless limited. This whole question should be studied in depth, with regard to socio-psychological as well as technological aspects, the ways of preventing illegal actions and the overall design of nuclear systems.

The importance of the problems of nuclear safety and radioactive waste management is further increased by the fact that public opinion is highly sensitive to these issues. The question of the public's attitude to nuclear energy is not however confined to the objective problems of its large-scale development. It is significant for example that other energy technologies, potentially dangerous for human health and for the environment, have never raised among the public the kind of reactions that have been occasioned by the hazards of nuclear technologies. This is partly due to the inappropriate association in the public's mind of civil nuclear developments with military uses of nuclear energy but more so to the lack of adequate and objective information made available to the public which is also often exposed to controversial experts' opinions. There is therefore a good case for undertaking research in social sciences so as to arrive at a better understanding of the factors determining public reaction to nuclear energy and other energy technologies. It would then be possible to develop means of objective information which are necessary if the development of nuclear energy is to be accepted.

III <u>OTHER ENERGY SOURCES</u>

Research and development of other primary sources of energy, such as the sun, the geothermal gradients, wind, tides, the ocean thermal gradients and currents and organic materials and waste, have until recently been mainly the concern of a few isolated individuals and groups within the scientific community. The interest shown so far by industry and governments has been limited either to small scale exploitation or to typical applications including for instance, in the case of France, the Odeillo solar furnace and La Rance tidal power station.

The oil crisis, however, and the various assessments recently made of the earth's energy resources, underlining their finite nature, have created a new awareness of the importance of all these forms of energy and therefore stimulated a new interest in their development.

Because of their great energy potential, most of these sources could in the future provide large and, in some cases, renewable supplies of energy; they constitute an interesting option in view of a policy of energy diversification. The R & D effort to develop these sources should in any case be guided not only by the concern for satisfying an increasing energy demand but also for providing a "quality" energy more adequate to the requirements of an environment-conscious society. It has also been underlined that these sources offer interesting opportunities for developing countries where some of them are particularly abundant (i.e. solar energy). A brief review of the prospects of application in those countries appears separately in Part III of this report.

This chapter describes many possible developments of these sources, whether for small scale local utilisation or widespread general use, and which range from applications foreseeable in the near future to projects now at the conceptual stage, the technological feasibility of which is unpredictable at present. <u>However, only those developments the implementation of which appears realistically promising have been mentioned in the summary of technological prospects included at the end of this report</u>.

In view of our relative lack of knowledge regarding the economic, and in some cases the technological, performance of these new sources, it is difficult to make realistic evaluations at present. This chapter is therefore to be considered illustrative rather than definitive where development prospects and cost estimates are concerned.

A. GEOTHERMAL ENERGY

The present geothermal electricity generating capacity in the world (Italy, United States, New Zealand, Mexico, Japan, USSR, Iceland) is a little more than 1,000 MW(1), comparable with the capacity of one modern nuclear power station. Other plants either planned or under construction in the world may well bring this figure up to 2000 MW by 1980. To this should be added the geothermal hot water produced and partly used at present in various countries for domestic heating, air conditioning, agriculture and industrial processing, roughly estimated to amount to 4,000 to 5,000 MW of heat equivalent(2).

It is evident from these figures that geothermal energy at present accounts for only a very small part of world energy supplies. Future prospects for its development depend on two factors: a) the amount of resources available and b) solving the complex technological problems associated with its large scale exploitation.

Because of the present insufficient knowledge of the geology and structure of the geothermal reservoirs as well as of their production capacity and depletion times, there is much uncertainty as to the amount of resources available and estimates made by various sources differ widely. World reserves have been estimated at 6,000 - 7,000 billion toe(3). For the United States alone, for instance, the geothermal electric potential of identified recoverable resources has been estimated to represent 1000 MW whilst a figure as high as 75,000,000 MW has been advanced; this includes undiscovered resources and implies major technological developments.(4)

A better assessment of the potential of geothermal deposits does in fact require a major research effort aimed at improving

---

1) -Exactly 1,010,850 KW at the end of 1972. Quoted in T. Leardini: Geothermal Power, Report presented at the Royal Society Meeting "Energy in the 1980s", London, 15th Nov. 1973.
- 1,000 MW = 3.15×10^{16} joule per year.

2) 4,000 MW = 12.6×10^{16} joule per year.
5,000 MW = 15.8×10^{16} joule per year.

3) 2.5×10^{23} - 2.9×10^{23} joule. Quoted in: Current Trends in Innovation in the Energy Sector, UN Economic and Social Council, Sc. Tech./R.9, 12th November, 1973.

4) 75,000,000 MW = 236×10^{19} joule per year. Quoted in: Energy Research and Development, Hearings before the Subcommittee on Science and Astronautics, House of Representatives, 92nd Congress, 2nd Session, May 1972.

detection and exploration techniques, as none of the present methods, including geophysical and geochemical techniques, and electrical measurements, appear to yield comprehensive results. For this reason it would seem very important to ensure that international co-operation in this field is effective and that the results of all research undertaken be made available to all countries. There is no doubt, for example, that if ideas of launching a geothermal satellite by NASA in the United States turn into a concrete project, the data that could be gathered by such a survey would be of great interest to most countries.

Forecasts of future developments of geothermal energy are too largely speculative. Estimates differ depending on the source and the level of the research effort proposed. For the United States alone, for instance, they range from 3,500 to 132,000 MW for 1985 and from 40,000 MW to 395,000 MW for the year 2000.(1) However, the technological problems which stand in the way of major development projects for the future are very acute.

It should be emphasized that although geothermal dry steam fields have been generating electricity for a number of years (mainly in Italy and in the United States) and geothermal hot water is used in several countries for house heating, these forms are most suitable for local applications. A more significant contribution by geothermal energy will largely be determined by the successful exploitation of other forms, such as hot brines and more particularly hot dry rocks, deposits of which are widely distributed and estimated to be ten times more abundant than steam and water fields. But power generation from hot brines, besides creating very serious pollution and environmental problems, requires a great technological effort and research in this field is still experimental. As to hot dry rocks, this form is a recent target of geothermal research and the technology for extracting heat to generate power is even less advanced.

It is therefore most important to distinguish between the different forms of geothermal energy and single out the different problems associated with each of them. Indeed, sources of heat - resulting from intrusion of the magma into the earth's crust or due to geological deformation - are unequally distributed with respect to their

---

1) Final Environmental Statement for the Geothermal Leasing Program, Vol. I of IV, United States Department of the Interior, 1973.

$3,500$ MW = 11×10^{16} joule per year.

$132,000$ MW = 416×10^{16} joule per year.

$40,000$ MW = 120×10^{16} joule per year.

$395,000$ MW = $1,240 \times 10^{16}$ joule per year.

thermal gradient and depth. There are consequently two main types of reservoir: low temperature and high temperature. In low temperature reservoirs, located mainly in sedimentary basins, hot water is found at a depth of 1,500 - 2,000 metres and at temperatures ranging from approximately 60°C to 120°C. The high temperature reservoirs of about 200°C to 350°C, are located in continental areas of relatively recent volcanic and tectonic activity and are found at depths ranging from a few hundred to several thousand metres. Here, depending on different conditions of temperature and pressure, heat can be found in the form of dry steam or hot brines. At those depths energy can also be found in the form of heat derived from hot rocks.

Hot Water

Hot water fields, containing water at low temperatures - from 60°C to 120°C - have been successfully exploited for purposes such as domestic heating, air conditioning, agriculture and industrial process heat. In Iceland, for instance, the city of Reykjavik is almost exclusively heated by geothermal water. In continental Europe hot water fields located in several sedimentary basins are found at depths not exceeding 2,000 metres. Thus in France about 2,000 house units in the Melun region have been heated by geothermal water since 1971 whilst new applications of hot water are being developed in other regions. In Hungary the land area covered by green-houses heated by geothermal water exceeded 1.5 million square metres in 1973.

Such fields are widely distributed in the earth's crust and could therefore be an important source of energy supply mainly for house heating, thus allowing for considerable fuel saving. The utilisation of geothermal hot water at such temperatures is, for the time being, restricted to this type of application. It is foreseeable, however, that part of this energy will be utilised in the future for power generation when more progress has been made in power conversion techniques.

Dry Steam

As has already been stated, it is relatively easy to exploit fields of dry steam, largely as a result of pioneering work in Italy at the Larderello plant. However, such fields are rare and the only one of significant size known so far, besides Larderello, is The Geysers field in California. Larderello and The Geysers (about 400 MW each at the end of 1973) have been harnessed for the generation of electricity since 1904 and 1960 respectively(1). Both construction and operating costs for the geothermal plants installed

1) 400 MW = 1.26×10^{16} joule per year

above these fields have proved to be lower than those for plants using fossil or nuclear fuel.

From a thermodynamic and economic point of view dry steam fields are certainly most suited to power generation, but the energy produced is not pollution-free, although this does not constitute a major problem. The hydrogen sulphide contained in steam is roughly equivalent to the sulphur released from burning low sulphur oil in a plant of comparable size.

Moreover, although the 30-40 years lifetime of steam reservoirs has indicated the economics of natural steam exploitation in some countries, if other fields are discovered, improved techniques allowing of the evaluation of the lifetime of steam reserves would encourage industry to make the necessary investment for their exploitation.

Hot Brines

Exploitation of hot brines for power generation raises very serious problems of corrosion and pollution as well as of water disposal. Briny water of 240-300°C emerges, as it flows up from the well, in the form of steam and water. With the exception of some areas (New Zealand, for instance) where the salinity is low, the water flowing with the steam generally contains a large amount of dissolved minerals (the Imperial Valley wells in California contain as much as 25 per cent minerals, compared to 3 per cent in seawater).

The high corrosiveness of such water is not only extremely harmful to the equipment but also constitutes one of the major obstacles to its use. No matter how briny water is due to be disposed of, i.e. whether treated in a desalination plant for the production of fresh water or re-injected into deep wells, handling, pumping and transporting it present real difficulties.

It is estimated that the quantity of water resulting from the generation of 1,000 MW(1) of electricity is as much as 570 million litres per day, which if evaporated would give 12,000 tons of minerals on the basis of 2 per cent salinity. This gives an idea of the magnitude of the problem and also points to the opportunity of associating the exploitation of hot brines for the generation of electricity with the production of minerals for the chemical industry.

Water could also be made available for agricultural and municipal use if it were suitably treated to remove the mineral content. But continuous removal of water from reservoirs can lead to subsidence, as has been the case in the Cerro Prieto fields in Mexico. Re-injection of water would thus have to be considered.

---

1) 1,000 MW = 3.15×10^{16} joule per year.

An alternative method is being investigated at Imperial Valley, California, where an attempt is being made to generate power from hot brines, keeping them under high pressure and transferring their heat to a liquid such as isobutane through a heat exchanger. Special alloys would be required to avoid corrosion in the heat exchanger. The brines, once passed through the heat exchanger, would be re-injected into deep wells, thus averting the risk of subsidence. If it proves successful and economically viable, this technology would make possible the utilisation of a greater portion of the heat contained in low and high temperature reservoirs for electricity generation.

Another disadvantage of hot brines is that the emission of hydrogen sulphide is higher than that of dry steam. Estimates indicate that the sulphur produced in a 1,000 MW plant would exceed the amount released from burning high sulphur fuel in a plant of comparable size.

No forecast can be made of the economics of geothermal power plants until more experiments are conducted to establish a proper basis of estimation. It is clear, however, that taking into account the costly process of re-injection or disposal of saline water, electricity generation from hot brines may be at an economic disadvantage compared with fossil fuel depending upon the cost of the ultimate fuel source.

Hot Dry Rocks

A new development in geothermal energy is the exploitation of the heat contained in hot dry rocks extensive deposits of which are at depths ranging from 2,000 to a maximum of 10,000 metres and at temperatures up to over 300°C. In areas with an abnormal thermal gradient (the average thermal gradient is 30°C/Km), it may be foreseen that rocks at 300°C will be found at only 2,000 - 3,000 metres.

A newly proposed method based on hydrofracturing techniques is to make large cracks in the rocks where pressurised water injected down from the surface would circulate, be heated and pumped up through another well for extraction of the heat. The project is under-way and it already appears that the hydro-fracturing techniques commonly employed for cracking oil-bearing formations are applicable to other rocks. At present this system appears more promising than the method of creating cavities by nuclear explosions in view of the higher cost of the latter and the possible danger of seismic effects.

At the experimental stage, the cost of one or two feasibility experiments is estimated at 20 million United States dollars, which shows how costly tapping the sizeable reserves of hot rocks might be.

It follows that if geothermal heat is to provide a large supply of energy, a major research effort will be necessary not only to pinpoint sources and geological anomalies but also to develop corrosion-resistant equipment and materials, to devise new drilling techniques suitable for hard and igneous rock at considerable depths, and to discover how the rocks will react to various temperatures and pressures. The solution of these problems will require extensive research and time and, for a number of years, the tapping of geothermal energy will be limited to local applications.

B. SOLAR ENERGY

Solar energy is the most abundant source of energy available to man. The small proportion of the sun's energy arriving on earth is many thousands times greater than the total capacity installed on earth in all the different forms. The average daily solar radiation for the United States is for example, equivalent to 3.7 thermal Kwh/mt^2 calculated on the basis of a 24 hour period and of the four seasons of the year. Thus, the energy arriving on 0.5 per cent of the territory of the United States of America would exceed United States energy needs projected to the year 2000.

The impact of the different applications of solar energy on energy consumption will depend on the level of R & D effort undertaken to develop this source. According to estimates, with a substantial development programme, solar energy could provide 35 per cent of total thermal needs in buildings, 30 per cent of gaseous fuel, 10 per cent of the liquid fuel and 20 per cent of electric energy requirements of the United States of America by the year 2020(1).

Research into solar energy is relatively advanced as compared with that on all sources other than conventional and nuclear ones. The full involvement of industry will be required if its widespread application is to be achieved in the future.

It is important however to distinguish between the different methods of using the sun's energy and to realise that, whereas water heating is a reality in many countries to day, power generation on a large scale is not expected to have any practical applications for the next 15 to 20 years.

The direct collection of solar radiations through technological means and indirect collection through the natural photosynthesis process offer a wide range of options of energy supply for water and house heating, power generation and synthetic fuels. Exploitation of the sun for these various applications is subject to one common

1) Solar Energy as a National Energy Resource, NSF/NASA Solar Energy Panel, Dec. 1972.

constraint - the intermittency and diffusion of sunlight. Clearly, however, the nature and scope of the problems is not the same for small-scale applications like water heating as it is for large-scale applications such as power generation.

1. Small-scale applications

The various technologies for small-scale applications are unevenly developed. Different technical approaches and improvements are needed in different cases. The main concerns in this field are the optimisation of systems and designs, engineering improvements and large-scale manufacturing of equipment in order to reduce costs.

Water heating, by means of roof collectors, is the most developed of the three thermal applications of solar energy to buildings. Roof collectors are at present commercially manufactured and used in sunny regions of several countries (Australia, France, Israel, Japan, United States, USSR). The availability of cheap and abundant fossil fuels has in the past discouraged the use of solar collectors for water heating but the increased prices and shortage of such fuels make this application of solar energy economically sound. However, a wider use of the system and its acceptance by the public will ultimately be made possible by the large-scale manufacture of collectors, which would allow of a further reduction in costs. Improvement of collector design and of absorbing surface coating could also help in this direction.

Space heating is less developed than water heating and is still in the experimental stage. Systems comprising different types of roof or vertical collectors have been tested and operated on experimental house units. Simple systems using concrete walls as vertical collectors are considered to be satisfactory on condition that an auxiliary energy system is available. Attempts to develop more sophisticated and autonomous types of heating envisaging storage of solar energy are considered so far to be too expensive and not particularly efficient. More R & D work is required in this field also in view of the fact that the efficiency of solar heating is related to the latitude and climatic conditions of the countries where it is to be applied. Likewise, the engineering aspects of collectors, storage systems and heat transfer units need further study.

As regards economic viability, experiments carried out in different United States climates have shown that solar heating is in every case competitive with and in fact cheaper than electric heating and, in a few cases even cheaper than heating by fossil fuels. Improved design and large-scale manufacturing would be required to reduce costs further.

Allowing for social-economic factors such as consumers' reluctance to accept higher initial construction costs despite lower operating costs, and the slow rate of new house building, much more than a decade will elapse before this form of heating can come into a more general use and affect total energy consumption.

Air conditioning is the least developed of the solar energy systems for domestic use. Few experiments have so far been carried out and substantial technical problems remain to be solved in the design of cooling systems. Some experts consider the absorption-refrigeration system as the most promising approach; this however, necessitates temperatures higher than those needed for heating. More research is therefore needed to improve collector efficiency by means of surface coatings. Cooling equipment and alternative storage methods also call for further research.

Special attention and considerable R & D should be devoted to combined heating and cooling systems but very little work has been done on these to date. As solar energy systems require capital-intensive equipment, combined systems of this kind would ensure better utilisation of such equipment and would thus reduce costs as compared with separate heating and cooling systems.

2. Large-scale applications

The technological problems and obstacles standing in the way of large-scale applications such as electricity generation are different from those referred to above, and they justify a more cautious approach to the various schemes that have been developed. Both terrestrial and space power generation have been envisaged and although it is unlikely that practical applications will be possible before the end of this century, in the long term the development of cheaper solar cells is considered as encouraging for photovoltaic conversion on earth.

i) Terrestrial Power generation

The two fundamental concepts of electricity generation from solar energy on the earth's surface are thermal conversion of sunlight by means of solar collectors and direct photovoltaic conversion in solar cells.

In the thermal conversion process, the heat collected by the optical concentrator and absorbed by the receiver is transferred to a turbogenerator. Studies of large, central generating stations based on this process are still in an embryonic stage. However, a number of substantial problems have already come to light.

The first set of difficulties is due to the size of the components required (mirrors or lenses and solar absorbers) which will have to be larger than any made so far, and to their durability and cost. The second set of problems relates to heat transfer and storage media. Some approaches appear satisfactory, such as heat transfer by nitrogen gas, liquid metal or a molten salt mixture, or storage by a eutectic mixture of salts, but here again the long distances from the collectors to the generating equipment, and the quantity of heat to be stored for a large plant, add greatly to the complexity of such methods. More research needs to be done on these processes in order to resolve the very important problem of heat transfer and storage.

Moreover, the economics of a solar power station, given the very high capital costs involved, are by no means comparable with those of conventional thermal or nuclear plants. Estimates indicate that solar power plants would be at least five to ten times more costly than conventional plants in view of the very high maintenance costs. Thus for a 1,000 MW peak capacity thermal power plant the investment required would be 2,000 to 2,500 million dollars and a mean kilowatt of capacity would cost $2,500. These are very tentative estimates and although their accuracy cannot be established until further engineering studies are undertaken on materials, and until the key technological factors have been tested, it appears already that thermal conversion of solar energy will be appropriate only for special applications. This will exclude its generalisation.

Photovoltaic conversion in solar cells is another proposed method of using solar energy to generate electricity. Like the solar collectors designed for solar heating units, arrays of solar cells would collect the sun's light, but in this process they would convert it directly, without an intermediate thermodynamic cycle.

Although this technology is in principle well known and valid since it has been practised in space exploration, the main difficulty up to now has been economic. The cost of single crystal silicon solar cells, manufactured in small quantities and assembled by hand, is extremely high and must be reduced by a factor of one hundred for the process to be economically viable. Investigations at present under way on the possibility of developing thin films of polycrystalline material, including silicon, cadmium sulphide and gallium arsenide, to replace single crystal short lived cells, justify some optimism for the realisation, in the long term, of cheap cells for photovoltaic conversion.

Even with an efficiency as low as 4 per cent, thin films would make photovoltaic conversion economically viable if mass production were developed. Whilst research to demonstrate technical feasibility should continue new methods of manufacturing cheaper long-lived cells must also be devised. Present research plans attempt a reduction in cost of solar cells through the development of thin films of polycrystalline material which would bring, in the next decade, the kilowatt production by solar energy to much the same level as that by conventional and nuclear fuels.

A main problem connected with photovoltaic conversion is storage as this process produces electricity directly. Storage in advanced batteries and flywheels would constitute an advantageous, economic method of ensuring low-cost, efficient storage although no practical systems have yet been developed(1).

ii) Power generation in space

A more speculative scheme proposed for power generation is that which consists of putting panels of solar cells into synchronous orbit. Electricity generated by these solar cells would be converted into microwave power, beamed to a giant antenna on earth and then converted back into electricity.

In view of the enormous technological problems and the prohibitive economics of such a scheme as well as of the ultimate danger of microwave radiations on earth, the generation of electricity by means of solar cells in space will for a long time be confined to spacecraft.

o
o o

Any assessment of the solar energy potential must take into account not only the resources resulting from the technological conversion of sunlight, as described above, but also those likely to be made available from the natural collection of solar radiations in the photosynthesis process.

Although this is an indirect application of solar energy, the conversion of photosynthetic production has been included in the following section together with conversion of waste because of the similarity of the scientific principles involved.

C. FUELS FROM ORGANIC MATTER AND WASTE

Research on conversion of organic materials and waste into synthetic fuels is of particular importance today in view of the partly renewable nature of these sources and because such conversion might represent a solution to the serious environmental problem of waste disposal.

Plants, trees and algae, can be grown for the ultimate purpose of providing heat through direct combustion or for producing synthetic fuels through biological and chemical conversions. Similarly, organic waste such as urban refuse, agricultural crop residues, animal and industrial waste, can provide heat or fuels when converted by the same processes.

Although available estimates are dispersed and fragmentary, the extent of the resources appears significant and justifies increased support for further research and development.

1) Cf. Chapter IV, Part II

1. Production of organic matter from photosynthesis

Land and water plants, grasses and algae represent a rich source of raw material which if converted into oil, methane or even ethyl alcohol could satisfy some of the energy requirements. Expert estimates suggest that the cultivation of plants on 3 per cent of the United States land surface would produce enough heat equivalent to cover the anticipated United States consumption of electricity in 1985. But these estimates are based on the assumption that the plant-solar conversion rate, which for most plants averages at present 1 per cent, would be increased to more than 3 per cent. Absorption of solar radiations is higher for some land plants, such as sugarcane and corn, and for water plants and algae. Compared to land plants, algae have a much higher heat value because of their higher protein and fat content and if converted into methane they could make a large contribution to the energy supply. According to experts such a contribution might be sufficient to cover total United States gas needs in 2020 by cultivating only 5 per cent of the United States land surface.

The low solar conversion efficiency of most plants is one of the problems where research is required to reduce the amount of raw material per BTU of fuel. Efforts should also be directed to improving plant productivity which is still too low, particularly for agricultural crops and trees whilst water plants, mainly algae, have a plant productivity which is five to ten times higher. Other important problems associated with photosynthetic production of organic materials are the high cost of land, and of plant cultivation, harvesting and transportation.

Because photosynthesis is viewed today as an important process for the purpose of supplying raw energy material rather than other products, new methods in plant genetics should be applied to all plants and particularly to plant species with high heat value, and improved agricultural processes should be developed to obtain large quantities of raw materials. Research in these fields should be coupled to the development of improved cheap procedures for cultivation, harvesting and transportation and the identification of low cost land to bring about reduction in the high cost of these items.

2. Organic waste

Contrary to land and water plants which have to be grown purposely for the production of fuel, waste including urban refuse, agricultural, animal and industrial waste, is available in industrialised societies in increasing quantities and has to be in any event collected, transported and disposed of without causing damage to the environment.

The heat equivalent of organic waste has been the object of several estimates. In the United States, for instance, waste production has been estimated to provide as much as 4 per cent of the annual United States energy consumption on the assumption that only about 50 per cent of it is collectable. Other calculations made on readily collectable waste in 1971 (only 15 per cent of the total) suggest an energy equivalent amounting to 3 per cent of oil or 6 per cent of gas consumption in the United States. In electric power equivalent, estimates of waste potential range from 6 per cent to 10 per cent of present electricity requirements. Thus, the contribution of this source, if not a major one, could be significant especially for heavily populated urban and agricultural regions.

The most important incentive for recycling waste is not so much to contribute quantitatively to the energy supply as to produce energy whilst providing solutions to some of the environmental problems associated with highly developed economies.

3. Conversion processes

Conversion of both the photosynthetic production and of organic waste into fuel appears today very promising and the technology for extracting oil, methane or other fuels from these two sources is progressing rapidly.

Different chemical and biological conversion processes are used depending on the type of raw material to be converted, its water content and its uniformity. The three main ones which appear to offer the best and most economical option for fuel production are pyrolysis, hydrogenation and anaerobic fermentation.

a) Pyrolysis

By this process, called "destructive distillation", organic materials are shredded and heated in a closed vessel, in an oxygen free atmosphere, at temperatures ranging from 500° to 900°C. Given the present state of technology, pyrolysis appears to be a most promising technique which is at the pilot plant stage. Conversion by pyrolysis can produce gas, oil and char but an advanced integrated system for processing urban solid waste allows also the recovery of ferrous metals and glass in addition to oil and char. Oil produced from urban waste with this system has been estimated to cost $8.4 per barrel in a 1,000 ton per day plant, whilst the estimated cost per barrel rises to $16.6 for the conversion of purposely grown organic matter. Such costs would diminish however in a larger scale production plant.(1) The products recovered from such a process also need to be upgraded and further research would be necessary in this field for such materials to be easily marketed.

b) Hydrogenation

Hydrogenation, or chemical reduction, is a less advanced process than pyrolysis. By this process, organic materials are heated at high temperatures (300° to 350°C) and pressures in the presence of water, carbon monoxide and catalyst. A heavy paraffinic oil(2) has been produced in a laboratory scale plant at a rate of 1.25 barrels (200 litres) per ton of dry waste.(3) At the pilot plant stage the value of such a product, estimated at around $4 to $5 per barrel, appears to justify production of synthetic fuels from waste, taking into account the cost of incineration waste disposal methods averaging $10 per ton. Economic feasibility of this process is however to be more accurately evaluated because many technical problems are still to be solved and because of the difficulty of assessing at present the operating cost of a commercial size plant.

c) Biological conversion by anaerobic fermentation

Anaerobic fermentation is a relatively simple process whereby organic materials are placed in a moist environment in an oxygen free fermentation tank equipped with a stirring mechanism. The biological conversion of organic materials through anaerobic fermentation for the primary purpose of producing fuel, and particularly methane, is still in its infancy, although this process is widely used in sewage treatment plants. Many technical and environmental problems remain to be solved, in particular the disposal of the high amount of sludge produced with this process. Feasibility studies covering investigation of the bacterial process, maximisation of fuel production and environmental effects are under way but it would be difficult to forecast when practical applications will be possible before this process has been demonstrated in a pilot plant.

---

1) Conversion into barrel has been calculated on the basis of one barrel of crude = 5.8 millions BTU.
   $8.4 per barrel = $14.2 per 10^{10} joule
   $16.6 per barrel = $28.1 per 10^{10} joule

2) (10 per cent oxygen content and 5 per cent nitrogen content as compared to 2 per cent of combined oxygen and nitrogen content in the widely used N°. 6 fuel oil).

3) 1.25 barrels = 0.74 x 10^{10} joule

The preliminary cost estimates for methane production range approximately from $8.7 to $11.6 per barrel when converting algae and purposely grown and harvested plants(1). Methane produced from urban waste is however estimated to be competitive with imported liquefied natural gas.

No significant contribution from photosynthetic production is expected for fifteen to twenty years and considerable research is needed to achieve efficient conversion into fuels and to improve agriculture and management techniques. According to the estimates now available, fuels produced from purposely grown crops are not competitive and innovative methods have to be developed to bring about the necessary cost reductions.

As more experiments have been made on waste conversion and pilot plants have been operating for some time, waste processing might be ready for commercialisation well before. Fuels produced by conversion of waste are equally not competitive but taking into account the cost of conventional incineration methods, they would be interesting side-products of waste disposal.

D. ENERGY FROM WIND, TIDES, OCEAN THERMAL GRADIENTS AND OCEAN CURRENTS

Winds, tides, ocean thermal gradients and ocean currents are sources of renewable energy supplies which are largely undeveloped at present, although the technical feasibility of some of them has already been demonstrated.

Wind power, for instance, was already harnessed a few decades ago to generate electricity, but its development for large scale applications is limited by several factors. First, this source is an intermittent one which makes electricity generation unpredictable and liable to interruptions and this necessitates important storage means. Other major obstacles are the difficult engineering problems involved, the high capital costs and the limited availability of accessible sites.

Nevertheless, this source can have very useful applications for pumping water or generating electricity, in small capacity installations, in isolated and windy regions. Programmes aiming at the development of wind power generating systems of 5 to 10 MW are at present foreseen and the new designs envisaged for such systems as wind turbines might allow of the economic generation of electricity for small scale local application.

---

1) Conversion into barrel has been calculated on the basis of one barrel of crude = 5.8 millions BTU.
$8.7 per barrel = $14.1 per 10^{10} joule
$11.6 per barrel = $19.6 per 10^{10} joule

Tides throughout the world have a potential as an energy source but this has been estimated to be not more than 2 per cent of the world's total hydro potential. The technological feasibility of tidal power stations has been demonstrated, and a 342 MW unit has been operating in particularly good tide conditions at La Rance in France for several years. Nonetheless, the high capital cost, the difficulty of practical development and the possibility of environmental effects on coastal areas discourage a major development effort of this source.

Ocean thermal gradients. A few conceptual designs for power generation based on ocean thermal gradients have been proposed by scientists and are at present being considered with a view to perfecting the design of a 100 MW floating power plant. The engineering evaluation which will result from such a study will indicate how energy from ocean gradients can best be utilised. Theoretically, it is thought that electricity generated through such a scheme should be used to electrolyse water. Whilst the oxygen could be returned to the sea, the hydrogen produced could be a considerable source of fuel that could be easily transported to the shore. No heat collector or storage unit, which are the most expensive and technically complex components for solar power plants, would be required for this kind of plant.

Ocean gradients might provide a renewable source of energy but their exploitation is limited to tropical seas with relatively high surfare temperatures. Any attempt to assess the technical feasibility of such a conversion process is of course impossible at this stage and in any event it is still an open question whether it is economically viable.

Ocean currents do not appear likely to be exploited for power generation on account of both the great technical difficulties involved and the prohibitive capital investment required. This energy source would be limited to areas with strong and fast currents and there are no prospects at present for its future development. Although proposals have been made for the exploitation of ocean currents in countries where their energy potential is high, it is recognised that the necessary research and development would be extremely expensive.

Although it is true that all these sources have a high energy potential, it is too early to determine what their contribution could be and whether they can be exploited economically. Some of them appear nevertheless to be fruitful areas of research.

IV ENERGY CARRIERS AND RELATED TECHNOLOGIES

A. REMARKS

Most of the energy resources discussed in the preceding chapters cannot be utilised directly because the process of producing primary energy from them differs fundamentally from the way in which the energy is finally used. Schematically, the differences relate to two factors:

i) Geographical pattern - The production of primary energy is concentrated in relatively large units: for example, oilfields, coal mines, nuclear reactors and geothermal sources, whereas the energy is finally consumed in units - many of them very small - which are scattered throughout the various countries.

ii) Time structure - The production of primary energy is generally a continuous process in which any changes take place relatively slowly, while the final demand for energy to be met at any time changes continually over a wide range with its own daily and seasonal variations.

From the resource stage to that of final utilisation, energy thus has to be transported and stored (i.e. transport in time). The problem is that energy, when finally used, is generally in the form of heat and mechanical energy, both highly unsuitable for storage and even less so for transport. Natural gas is the only energy resource that can be transported, stored and used just as it is for a large number of purposes.

All other sources of primary energy must therefore be converted into secondary energy agents, or energy carriers, distributed to the points where they are put to various uses. Energy carriers are key elements in energy systems and are of fundamental importance; they determine whether the many kinds of resource reviewed in the preceding chapters can be used effectively; they should therefore be made the subject of an appreciably intensified R & D effort.

The ideal energy carrier should have the following properties:

- it should be capable of being produced economically with a high efficiency from the various energy sources without causing environmental damage;

- it should be capable of economically transporting large amounts of energy over long distances without any loss and without in any way damaging the environment;
- it should be capable of storing economically both large and small amounts of energy, without any loss and without in any way damaging the environment;
- it should be capable of being used economically and with a high efficiency for a large number of purposes (industry, household and commercial applications, transport) without in any way damaging the environment.

It is clear, unfortunately, that no single energy carrier combines all these advantages. It will be necessary, for example, to have recourse to various energy carriers, adapted to different ends, and a compromise will have to be found between their advantages and drawbacks. In other words, the goal of research in this field will be not so much to find the ideal energy carrier as to optimise the whole system of energy carriers from the standpoint of economics, the environment and the efficient use of energy resources.

One of the main reasons for the importance that hydrocarbons have now acquired is their many advantages as energy carriers from the production, transport and storage aspects. This fact was mentioned earlier with regard to natural gas. Petroleum products (gasoline and fuel oil) are even better where transport and storage are concerned, since they are liquids with a relatively high specific energy; the technology of producing them by refining crude oil is already highly developed. One of the main goals of research into the gasification and liquefaction of coal mentioned in Chapter I is to obtain from this source the same energy carriers as those obtained from natural gas or crude oil.

On the other hand, fossil fuels have certain disadvantages as energy carriers when it comes to utilisation. First of all, directly they can produce only heat which is then converted very inefficiently into mechanical energy. Their utilisation moreover gives rise to serious environmental problems: besides the disadvantages caused by the presence of impurities,(1) mainly sulphur, in most oil (and especially in coal), the combustion of all fossil fuels generates carbon dioxide. While no ecological imbalance caused by amounts of carbon dioxide so far produced in this way has yet been detected on a world scale, it is by no means certain whether such an imbalance may not rapidly develop if the utilisation of fossil fuel goes on increasing exponentially, as it has so far done: this is a sufficiently serious problem to warrant thorough investigation.

---

1) Cf. Chapter I, Part II.

It is precisely because electricity offers so many advantages at the utilisation stage that it has developed as an energy carrier more rapidly than other secondary agents. This trend will probably be reinforced during the next few years as a result of the accelerated development of nuclear energy, which at present can only be used for generating electricity. This should not however disguise the fact that the generation, storage and transmission of electricity creates many problems which call for special attention; they also warrant an appreciable increase in R & D effort on other energy carriers.

B. ELECTRICITY

As electricity as such, economically, cannot now be stored in large quantities, the necessary continuous matching of supply to demand is effected by combining different kinds of power stations in the same generating system.(1) Schematically, the base load is covered mainly by large modern thermal power stations which have a relatively high efficiency and represent a substantial capital investment which must be made to yield a maximum return by running the plant on a continuous basis. Older, less efficient (but already amortized) thermal power stations, and hydro plants working from reservoirs, are normally brought in to cover daily variations in the load diagram, while peaks are covered by pumped-storage plants and gas turbine stations. This last type of plant has lower capital cost but also much lower efficiency.

1. Electricity generation

Disregarding hydro power stations which will become relatively less important as suitable sites become scarce, the process of converting the different primary energy agents into electricity (fossil fuels, nuclear and geothermal energy) is an indirect one: the primary energy agent produces heat which is transformed into mechanical energy, which in turn is converted into electricity.

It should be borne in mind that the efficiencies reached at this latter stage are already high. Furthermore, the development of superconducting alternators should make it possible to reach unit ratings appreciably higher than present limits and thus provide increased economies of scale. According to the present state of research in this field, this type of alternator could be gradually introduced into power stations during the next decade.

---

1) This question is discussed at greater length in the Report of the ad hoc Group of the Energy Committee on Electricity, OECD, (mimeographed).

The main problem is the conversion of the heat into mechanical energy. The thermodynamic efficiencies of base load power stations, i.e. the most efficient plants, amount to some 40 per cent in modern conventional stations and 32 per cent in existing nuclear plants. This means that nearly two-thirds of the heat produced is dissipated into the environment, mainly into watercourses. Such thermal pollution is the leading environmental problem when developing electricity supplies owing to the limited water resources, and their distribution, in most OECD countries.

The possible ways of resolving this problem fall under two headings and should be considered complementary: remedies and preventive action. One remedy would be to site power stations by the sea (or out at sea), since the ocean is a considerable source of cooling water; apart from the fact that this solution clearly cannot be applied to countries which have no seaboard, it would substantially add to the problem of transmitting electricity (to be discussed below). Another method of the same type is to use cooling towers to dissipate waste heat into the air instead of into watercourses. Quite apart from the aesthetic problems involved, it has been estimated that the capital cost of power stations would then increase by some 3 to 10 per cent and operating costs by some 6 to 14 per cent.

Generally speaking, remedial methods do not resolve the real problem which is poor utilisation of energy resources, thermal pollution being one consequence. R & D in this field should therefore be concentrated on "preventive" action, i.e. on ways of improving the energy efficiency of power stations,(1) which are discussed below.

One line of research is to <u>improve the thermodynamic cycles</u> used. Although it seems fairly unlikely that the steam turbines now in common use can be improved to any major extent, the development of high-temperature gas turbines and combined cycles should lead to significant progress. The development of high temperature gas turbines, which depends primarily on the development of suitable materials, is particularly attractive when considered together with that of high temperature gas-cooled reactors. The use of helium in a direct cycle for cooling these reactors would provide a relatively economic solution to the question of dry cooling. Another interesting possibility would be to employ a combined cycle, using the heat from the gas turbine exhaust gases in a "bottoming" steam cycle; the efficiency of the whole system could then reach 44 to 45 per cent.

---

1) It must however be added that research into "remedial" methods is worth continuing, since the progress to be expected from preventive methods will be so gradual and limited that no complete solution to the problem of thermal pollution will be forthcoming.

Combined cycles also appear necessary for using the low BTU gas obtained from coal conversion.(1) They could also be used in peak power generating plants where gas turbines use heat from fossil fuels; it has been estimated that in this way the efficiency of these power stations could be increased from about 24 per cent to nearly 38 per cent, but there would be a corresponding increase of about 60 per cent in capital costs.

A second line or research is aimed at the direct conversion of heat into electricity using magnetohydrodynamics (MHD) (i.e. producing electricity by passing an ionized gas through a magnetic field), which in theory could lead to efficiencies of 50 to 60 per cent. Of the three main types of MHD system - open cycle, closed cycle plasma and closed cycle liquid metal - the first is the most advanced, but it seems unlikely that MHD will be developed for the industrial generation of electricity before the end of the 1980s. One of the basic problems still to be resolved is that of the useful life of the materials from which the various components of MHD generators are made; it is therefore likely that in the first place MHD will be used for peak power generation.

Another method of appreciably improving the overall energy efficiency of electricity generating stations is to make use of their waste heat instead of dissipating it into the environment.(2) We would then no longer have a mere power station but a plant for converting primary energy agents into two energy carriers: heat and electricity. This method has already proved valuable in what are known as "total energy systems", i.e. consumption units requiring both heat and electricity (for example, hospitals, large shopping centres, and certain industries such as sugar refining). It can also be envisaged for the desalination of sea water.

In order to extend this method to large power stations and use their waste heat for such geographically scattered purposes as district heating or industrial and agricultural processes, intensified R & D in the field of heat transport is particularly needed. The temperature of residual heat from power stations is relatively low and a substantial proportion of its thermodynamic potential is lost after transmission over a few kilometres. The longest distances over which this method has been used for district heating reach some 40 kilometres. These distances need to be substantially increased if the method is to be properly developed. Research is also necessary into the management of these systems, which produce two energy carriers at the same time; in particular it will be important to resolve the problems arising from the considerable differences between daily and especially seasonal variations in demand for

1) Cf. Chapter I, Part II.

2) Cf. the above mentioned report of the ad hoc Group of the Energy Committee on Electricity.

electricity and that for heat. In any case, the use of waste heat from power plants represents a very important element in the problem of the siting of power plants as it requires these plants to be located near the heat utilisation centres.(1)

A fourth line of research is to study fuel cells, i.e. the direct production of electricity from the chemical energy of such fuels as hydrogen, alcohols, carbon monoxide or hydrocarbons. The main advantage of fuel cells would be their high efficiency (50 to 70 per cent), their quietness and great flexibility in use. Above all, they would make it possible to combine the advantages of electricity at the energy utilisation stage with those of liquid and gaseous energy carriers as regards transport and storage.

A substantial R & D effort would still seem necessary, particularly in electrochemistry and materials science, before fuel cells with sufficiently low capital cost, adequate working life and using relatively cheap fuels can be used for generating electricity other than in very special circumstances (for example in space applications). The best results have so far been obtained with hydrogen cells, but their use depends on the possibility of producing hydrogen economically; this point will be examined in more detail below. A similar remark applies to the methanol-air cell which could give equally interesting results. Moreover, these fuel cells have a limited power and need very costly catalysts such as platinum. In view of the extent of the technical and economic problems yet to be resolved, it is unlikely that fuel cells will be applied on a large scale before the middle of the next decade.

2. Storing electricity

The main drawback of electricity as an energy carrier is that it cannot be economically stored as such. This is why peak loads must be covered to a large extent by generating facilities such as gas turbines; these can be brought into service very quickly but have very low efficiency. Another method is to convert surplus electricity, generated during off-peak hours by more efficient stations, into some storable form, then to reconvert this energy into electricity at peak times. The advantage here is that the overall energy efficiency of the electricity generating system would thereby be increased.

One widely used method of storage consists in converting electricity into potential energy by pumping water into a high reservoir. The efficiency of the system is 68 per cent. The development of pumped storage schemes is however hampered by the shortage of suitable sites and by their very high capital cost.

1) In this respect, it would be important to have a further knowledge of geographical distribution of the needs for heat.

Three methods of storage which were recently suggested deserve careful study before substantial research programmes are devoted to them. The first would be to utilise any surplus electricity generated during off-peak hours for accumulating compressed air in underground excavations; the air would subsequently be injected into gas turbines, increasing their efficiency by a factor of between 2 and 3. The second method would be to convert electrical energy into mechanical energy which would be stored in flywheels and later reconverted into electricity; the use of this method, which is based upon a very old principle, would be possible owing to the recent development of materials such as carbon fibres which have high tensile strength. Another method envisages storing electricity in the form of hydrogen which would be produced by electrolysis.

A method of storing electric power which is at present the subject of an important amount of research is storage in secondary batteries. Batteries are not only one of the ways which can be envisaged for providing electricity during peak hours but also the principal requirement in the development of electric vehicles. Although the performances needed are not exactly the same for these two applications, many of the characteristics sought, in particular, low cost per kWh and high specific power (above 100 Wh/kg) are similar in both cases. The progress made during the last few years justifies a sustained R & D effort and suggests that there will be steady development during the 1980s. Batteries which seem of particular interest are zinc-air cells for low-temperature use and lithium-sulphur, sodium-sulphur and lithium-chlorine cells for high-temperature applications (operating between 300°C and 700°C). As in the case of fuel cells, electrochemical and materials research is necessary (in particular catalysts and electrodes), with a view to increasing the useful life and the number of charge-discharge cycles of these batteries.

Another possible method of storing electric power involves the use of large DC superconducting magnets. While for utility application it would be desirable to store between 10^{11} to 10^{14} joule in such inductors, superconducting magnets storing nearly 10^{9} joule have already been constructed for other purposes. This type of storage promises to be the most efficient of all types considered, approaching 95% efficiency. This figure includes the refrigeration energy needed to maintain the superconducting state as well as losses encountered in the AC-to-DC and DC-to-AC converters employed to interface the storage unit with the electrical power transmission system. Magnetic energy storage devices could be compact (energy densities of up to 90×10^{6} joule/m^{3} can be obtained with magnetic fields up to 15T), easily sitable near the load centres (probably underground), and would incur minimal environmental objections.

Considerable R & D is still required to establish optimum magnet and conductor configurations and mechanical support structures for containing the large magnetic forces. In addition, it is important to provide continuing support to research aiming at discovering superconducting materials with critical temperatures substantially higher than the temperatures of liquid helium and liquid hydrogen.

3. The transmission of electricity

As electricity is developed, the problem of its transmission will moreover become appreciably more acute. Electricity is at present transmitted by overhead lines, except in towns where it is distributed by underground cables. It must however be pointed out that power stations are now located near the main centres of consumption and that current is transmitted in large amounts only over fairly short distances averaging some 100 kilometres. While interconnecting national and international networks are of course much more extensive (covering several thousand kilometres), their purpose is not so much to carry large amounts of electricity as to permit an overall balance between generation and demand within a country or group of countries and as far as possible to compensate for any local and cyclical variations in generating capacity and consumption.

As stated above, the problem of siting power stations is becoming gradually more difficult, particularly owing to the question of water resources, and it may become necessary to site them farther from centres of consumption. One leading R & D objective should therefore be to find ways of transmitting amounts of electricity over longer and longer distances under conditions which are economically as well as environmentally adequate.

To keep losses sufficiently low (less than 5 per cent), while limiting the increase in numbers of lines, the first step is to increase the line voltage; the highest voltage now in use is 765 kV. Although some useful progress can be made in this direction, it is most unlikely that it will be possible to go beyond 3,000 kV, owing to such technical limitations, for example as those due to corona effects. DC lines are another possible means of transmitting large amounts of electricity, but it must not be forgotten that their development depends in particular on a sustained programme of R & D dealing with DC/AC conversion.

Because of the large amounts of space taken up by overhead lines, it will probably be necessary to turn more and more to underground cables, but their cost is such that this change will probably take place very gradually, beginning in the densest urban areas. The capital cost of present-day underground cables is about 15 times that of overhead lines and their performance is limited by problems of insulation and especially of removing the heat due to resistive losses. Many types of high-performance cable are under development, a most promising example being insulated by a compressed gas (sulphur hexafluoride). These cables could be gradually introduced from 1980 onwards. There are also cryoresistive cables operating at the temperature of liquid nitrogen; their introduction on the market, however, would not seem to be immediate.

In the longer term, superconducting cables seem capable of carrying very large amounts of energy, with virtually zero losses from direct-current cables and very low although non-zero losses A.C. operation. The amount of energy needed for cooling purposes and for circulating the refrigerant could be kept below 0.3 per cent of the transmitted power. Although most of the superconductivity technology involved is already at hand, R & D must be continued with respect to the economic and reliability aspects. Testing of the reliability of refrigeration systems and cryogenic envelopes are other areas requiring important R & D support. As in the case of electricity storage through superconducting magnets, the development of superconductors with higher critical temperatures represents a very important research objective.

C. OTHER ENERGY CARRIERS

In spite of the many advantages of electricity at the stage of final utilisation, its development, as we have seen, creates extremely difficult problems. It is therefore important to devote a very substantial R & D effort to other energy carriers lending themselves to the many applications where the specific properties of electricity are not needed, including all thermal applications, which account for a large proportion of final energy consumption.

In this connection, mention has been made of the use of the residual heat from electricity generating stations and the direct use of heat produced by high-temperature nuclear reactors. The possibilities of storing and transporting heat are however much too limited for development as an energy carrier to be considered other than in very localised applications (for example, district heating or industrial estates).(1) It is therefore necessary to look for liquid or gaseous energy carriers which can be obtained from abundant resources (other than natural hydrocarbons) and from heat (especially heat from nuclear reactors) and then restore energy in the form of heat at the final consumption stage. This is the objective for example, of the research already mentioned concerning gasification

1) Even for these localised applications, research is needed on the transport and storage of heat (for example, in the form of latent heat in molten salts).

and liquefaction of coal,(1) or the conversion of organic wastes into methane.(2) This is also the aim of research on methanol, hydrogen and more generally on the conversion of heat into chemical energy.

1. Methanol

Methanol is an alcohol which has been known for some time, but its use as an energy carrier is a more recent idea. It can be obtained from many organic materials: natural gas, coal, wood, other plants and organic wastes. Its calorific value is about half that of gasoline. At present, the main prospect for methanol is to use it as a substitute for gasoline. Methanol can be added to gasoline in proportions as yet uncertain without requiring any change to existing engines. It is also possible to use pure methanol or a mixture of methanol and water to fuel engines much like those we have today with less pollution. R & D on this subject is however still at an early stage and it is important that research into possible ways of using methanol be intensified.

The main advantage is that it can be transported and stored in facilities already used for oil, while the latter's disadvantages in the matter of water pollution and the safety problems of LNG can be avoided.(3)

For this reason, consideration has been given to transporting methanol instead of LNG by sea. Energy efficiency when methane is converted into methanol is however only 50 per cent and at present the process costs appreciably more than that of liquefying natural gas, which means that methanol is economically advantageous only when carried over sufficiently long distances.(4) It would now seem more expensive to produce methanol from coal than to convert coal into other fuels such as methane. These data should however not be taken as final, and one of the main aims of R & D in this field should accordingly be to reduce the cost of producing methanol from resources such as coal, e.g. through process heat from high temperature gas cooled nuclear reactors.

Furthermore, research should also deal with the possible production of methanol from wood and other plant life or from solid wastes.(5) It should be borne in mind that the economics of

1) See Chapter I, Part II.

2) See Chapter III, Part II.

3) Mention should be made however of the high toxicity of methanol.

4) The cost of energy now being produced from Algerian natural gas is virtually the same whether it is transported to the United States as methanol or liquid methane.

5) Cf. Chapter III, Part II.

producing methanol from solid wastes should be considered not only in relation to the economics of other energy carriers but also in conjunction with the general problem of waste management.

To summarise, the prospects held out by methanol are sufficiently attractive to warrant increased R & D possibly culminating in substantial progress as from 1980.

2. Hydrogen

As in the case of methanol the idea of using hydrogen as an energy carrier is a recent one. It must be remembered, however, that this can only be a long-term prospect owing to the difficult problems involved, in particular those connected with safety and the possibility of producing hydrogen economically.

Hydrogen is at present produced by catalytic cracking of hydrocarbons or by electrolysis of water. The disadvantage in the former process is that recourse must be made to resources which are already widely used as energy carriers, while in the second process the need for electric current leads to the inefficient and costly use of energy resources.(1) The purpose of most research on hydrogen production is therefore to discover a process enabling it to be obtained from water and heat alone. Since direct thermal dissociation of the water molecule calls for temperatures greater than 2,500°C and these cannot be obtained from primary energy resources, attempts are being made to dissociate the water by means of thermochemical cycles for which the necessary temperatures would be provided by future nuclear reactors, especially high-temperature reactors. Various cycles are now under investigation, but a substantial research effort is still needed if any really adequate cycle is to be developed. Another process of hydrogen production from water using the photosynthetic apparatus of green plants or algae is also being considered but it is still at the design stage. Extensive basic research is needed before the feasibility of biophotolysis and its economic viability can be established.

While no new technology is required for transmitting hydrogen (either gaseous or liquid), existing gas transmission facilities will have to be adapted, since hydrogen requires greater pumping capacity and is much more fluid than natural gas. As to storage, aquifers used for natural gas can also be used for hydrogen although the problems associated with the permeability of rocks are much more serious for hydrogen storage; moreover containers carrying up to 50,000 m^3 of liquefied hydrogen can be built without requiring any further basic development work.

---

1) It is however necessary to mention research on high temperature electrolysis.

Since gaseous hydrogen however requires much more storag
than natural gas, and liquid storage calls for costly cryogeni
containers, alternatives are being investigated, including sto
in the form of metal hydrides. While such metals as vanadium, nio-
bium and alloys such as the compound $LaNi_5$ when exposed to pressurised
hydrogen, can store it efficiently, their cost is very high. Less
expensive metals could make hydride storage feasible. New technology
is required for storing small quantities of hydrogen; thus metal-
air batteries, rechargeable with power from hydrogen, and titanium
hydride storage in tanks are possibilities which have been considered.

The cost of transmission for hydrogen is estimated to be some
two to three times higher than for natural gas but much lower than
for electricity, especially over long distances. Storage is also
two to three times more costly than for natural gas, but liquid
hydrogen can be stored at a fraction of the cost of storing electrical
energy in pumped-storage hydro schemes.

In principle, hydrogen has many applications. It can be used
as a fuel for domestic purposes or for motor vehicles and aircraft.
Compared with fossil fuels, it has the advantage that when burned it
produces water instead of carbon dioxide, although it must be pointed
out that various nitrogen oxides are produced in the case of combustion if hydrogen is burned in air. If hydrogen is burned in
oxygen, there is no production of nitrogen oxydes and the temperature
of combustion is much higher. A disadvantage of hydrogen is its relatively low calorific value per unit of volume (about one-third
that of methane); moreover, the problem of containers for limited
amounts of hydrogen which are light and small enough for transport
purposes is far from being resolved. Finally, hydrogen has some
marked disadvantages from the safety aspect when it comes into contact with the oxygen in the air.

Hydrogen alone would thus seem less attractive as an energy
carrier than when used in conjunction with other energy carriers.
In this connection, the two most promising approaches are its
possible use for the hydrogenation of coal in order to produce
liquid or gaseous hydrocarbons and its use in fuel cells for producing electricity. Hydrogen is of course also important for the
whole chemical industry.

While despite these drawbacks there is a good case for considerably increasing hydrogen research, the base should be widened
to include the transport of heat in the form of energy of the
chemical bond. In the case of hydrogen obtained from water, the
heat is converted in the form of energy resulting from dissociation
of the water molecule but other chemical reactions may also be
considered.

One well-known reaction of this type now being studied so that it can be used to transport energy is:

$$CH_4 + H_2O + 49\ \text{Kcal/mole} \rightleftarrows 3H_2 + CO.$$

At the production stage, conversion of the methane and water into hydrogen and carbon monoxide calls for a temperature of some 850°C, which can be supplied by nuclear reactors. The hydrogen and carbon monoxide can be carried by pipeline to the point of consumption, where they react by releasing their energy of the chemical bond, enabling the methane to be returned to the production plant. The advantages of this process would be transport over very long distances, reaction in a closed cycle with the result that no pollutants would be emitted, and the fact that only heat of nuclear origin and no hydrocarbons would be used (except for the initial charge of methane).

The system based upon the above reaction is not necessarily best suited for all applications, but it is a particularly apt illustration of the prospects offered by the general idea of transporting energy in the form of energy of the chemical bond. This idea warrants a substantial R & D effort, which would be to explore and evaluate any possibly useful chemical reactions and to tackle any technological and commercial difficulties which the most promising reactions might entail.

V UTILISATION AND CONSERVATION OF ENERGY

Preceding chapters have laid stress on the possibilities offered by science and technology for increasing the production of energy but quite as important a step is to direct research and development in such a way that demand will be reduced through a better utilisation of energy. It must be emphasized, however, that in trying to conserve natural energy resources, systems for producing and converting the energy extracted from them must first be dealt with; in this field of research it is important that the longer-term concern of optimising the energy yield of these systems should not be sacrificed to short-term economic considerations.

For several decades, the exceptionally low prices paid for oil have caused energy in the industrialised countries to be increasingly used and in some instances profligately so, although the progress made in industrial processes has sometimes led to a better utilisation of energy. As a result, there has been virtually no research aimed at improving the utilisation of energy. Only recently, owing to the difficulty of obtaining oil supplies and to the rising prices, has any consideration been given to expanding R & D in order to conserve energy at the stage of final utilisation.

The objective of energy conservation can be seen at different levels. Of course it can be to reduce the overall consumption of energy, whatever the sources from which such energy is produced. The various energy sources however are not equivalent; in particular some resources are more "scarce" and the aim may be to replace them by more abundant resources. For instance the trade balance problems associated with oil imports may lead to the substitution of nuclear electricity for oil products. In a similar way, the development of solar energy can be viewed as aiming at the conservation of other energy resources.

It must also be borne in mind that science and technology probably contribute less in this field than in the production of energy. The attitude of the many and varied types of consumer is at least as important a factor, and indeed the decisive factor in the short term. In this respect it should be noted that the present situation offers an unrivalled opportunity for experimental research in the social sciences, one which it would be well to seize for undertaking a detailed study of the effects economic or other incentives can have in reducing energy consumption, in promoting a lesser or more efficient use of motor vehicles, etc.

Despite these limitations, science and technology can significantly help in improving energy utilisation and R & D should hence be considerably intensified to this end. The problem is that so many different applications are involved that all the scientific and technological aspects are extremely hard to define,(1) and very careful consideration is needed before the directions R & D should take can be ascertained. A few general guidelines(2) dealing with the four main sectors of utilisation (household and commerce; agriculture; industry; transport) are given below.

The household and commercial sector

In this sector, the utilisation of energy could be improved appreciably by introducing a number of innovations, but it is mainly in space heating and air conditioning that substantial amounts of energy could be saved. It is in these applications that solar energy is likely to be of considerable help during the next ten years.

Among other possibilities offered by science and technology may be mentioned construction materials and techniques providing better thermal insulation and the increased efficiency of heating and air conditioning systems (for instance by making a more extensive use of heat pumps). Furthermore the architecture and overall design of buildings have a significant impact on energy consumption and should be carefully studied from this point of view.

It must however be pointed out that most technologies affecting this sector are already far advanced. Often the problem is not so much to improve their technical performance as to overcome the obstacles of all kinds which have so far stood in the way of their large-scale application; for example the way the construction industry is organised or the fact that these technologies call for increased capital expenditure.

In addition to technologies, the behaviour of energy users in the household and commercial sector should also be the subject of increased research. Examples of such research are the importance attached to energy consumption in the choice of domestic appliances and heating systems as well as in the utilisation of these appliances, the possibilities and limitations of replacing space heating and

1) In particular, one should not underestimate the importance of "microtechnologies"; their individual contribution may be limited but their combination may lead to a significant degree of energy conservation.

2) A more detailed list of possible research subjects will be found in the report Technology of Efficient Energy Utilisation, NATO, Brussels, 1974.

cooling by adapting clothing to temperature conditions, the real effectiveness of illumination for advertising purposes. Such research should make possible a better definition of the nature and modes of the actions undertaken to inform and educate users, whose behaviour will eventually determine to a large extent the degree of success of conservation policies in this sector.

Agriculture

Although energy saving policies have so far paid it little attention, agriculture in the developed countries is based to an increasing extent on the intensive use of energy, particularly that obtained from fossil fuels (this energy is added to the "natural" use of solar energy in the process of photosynthesis). The lack of precise data in energy consumption (by tractors, and in producing the different fertilizers and other products such as pesticides and herbicides) is even worse than in other sectors. This question would appear to merit a detailed study of itself. It has however been calculated that the annual consumption of energy by the agriculture industry was at least 4.7 Mtoe (197×10^{15} joule) in the United Kingdom,(1) and about 52.6 Mtoe (2.2×10^{18} joule) in the United States in 1970,(2) i.e. about 3.8 per cent of the total final consumption of energy.

These figures are even higher if we include the rest of the food system, i.e. the food industry and the storage and preparation of foodstuffs in the domestic sector. In the United States, the energy consumed by the food system in 1970 was of the order of 217.2 Mtoe (9.1×10^{18} joule),(2) i.e. more than 15.6 per cent of the total final consumption of energy.

This situation is the direct result of R & D spanning several decades and aimed at increasing the productivity of agriculture and the food system. The "Green Revolution", incidentally, is mainly a matter of introducing into the developing countries an agricultural system based upon the intensive use of energy comparable with that in developed countries.

Substantial energy savings could be achieved in the various agricultural sectors through a number of innovations. Using natural in place of chemical fertilizers would save about 0.04 toe per hectare and more frequent use of crop rotation would mean a saving of about 0.06 toe per hectare. The development of biological pesticides (e.g. those which sterilize male insects) would save much of the energy now being used to produce and apply chemical pesticides.

---

1) Dr. Kenneth Baxter, "Power and Agricultural Revolution", New Scientist, 14th February, 1974.

2) John S. Steinhart and Carol E. Steinhart, "Energy Use in the U.S. Food System", Science, 19th April, 1974.

Research could also be carried out on developing plant varieties that are more resistant to disease, more easily stored (in order to save energy used in their chemical treatment or storage) and with a higher protein content. In the longer term, research could also be directed to increasing the efficiency of the photosynthesis process which at present averages only 1 per cent.

If other elements of the food industry are considered, an R & D effort is also necessary to reduce energy consumption both in the food industry and at the domestic level. However these aspects are respectively part of the industrial sector and the domestic and commercial sector.

The industrial sector

Even before embarking on the subject of R & D, it should be realised that large energy savings could be achieved even with existing technologies, by means of suitable plant design, the wider use of automatic controls, proper maintenance of equipment, etc. Here again, the main problem consists in eliminating non-technical obstacles which have so far hampered the introduction of technologies or methods promoting the better use of energy.

While to a very large extent, the question of what research to undertake will have to be resolved by first accurately analysing the various industrial processes, a few general approaches which should prove important in many cases may be pointed out:

- Recovery of heat and elimination of heat losses. In this field, special attention should be given to the possibility of using large heat pumps, improving heat exchangers and improved insulation of furnaces and pipework.
- New methods of heating and improvement of existing techniques. For example, improved combustion techniques, improved induction heating, R.F. and microwave heating systems, laser or electron beam welding, etc. may be mentioned.
- Development of production methods reducing the wastage of materials with a high energy value to a minimum. For example, replacement of machining by methods such as powder metallurgy or plastic forming.
- Development of manufacturing processes requiring less energy. Research enabling present techniques of steel production (reduction of iron ore by coke in blast furnaces) to be replaced by processes such as the reduction of iron chloride by hydrogen is an example.
- Development of materials and products which have a lower energy value than materials now used, but with similar properties. In this respect it may be noted that determining the energy value of a material or a product is in itself a subject for research.

Transport

Research into energy conservation in the transport sector is of particular importance because the major portion of mechanical energy used for transport purposes is produced in engines which burn fuels derived from oil and which have a relatively low thermodynamic efficiency. The transport sector, moreover, consists of a wide variety of public and private elements providing a wide range of products and services. Effective research aimed at energy conservation can therefore be directed at the following three general areas: engineering improvements, legislative measures and institutional changes.

More specifically, engineering improvements include specific changes to the vehicle or control systems and the methods of construction and maintenance of the travelled way and terminals which reduce fuel consumption. Legislative measures involve fuel allocation schemes, vehicle acquisition and use restrictions, and vehicle and fuel taxation. Institutional factors include the complex set of relationships involving governmental structures, transport companies, labour unions, regulatory policies, subsidies and international agreements that set the framework in which the various modes of transport operate. Measures designed to alter the existing equilibrium of factors within any or all of these areas could bring about a saving of energy, but must be taken only with a full understanding of the consequences.

Engineering improvements, for example highway transport, include strategies for reducing vehicular fuel consumption with improved highway design features and with improved highway traffic control measures, alternative vehicle component and powerplant design and energy efficient highway construction and maintenance methods. Investigations into other forms and modes of people and goods transport may also yield significant fuel consumption reductions.

Among other engineering improvements, mention might also be made of research for improving the diesel engine, which is more efficient than the spark ignition internal combustion engine but which still has certain disadvantages (noise, lower power density, poorer acceleration, higher capital cost). Another promising approach seems to be the Stirling engine: this would have an efficiency matching that of the diesel engine over the whole power range, would be very quiet, cause very little pollution and be adaptable to the most varied kinds of fuel.

A careful examination of the kinds of travel, by mode and by purpose, should provide some indication of where substantial energy conservation can be expected and also provide a basis of longer range transport planning and development. For example, in urban areas in the United States, movement of people accounts for about 45 per cent of the fuel consumed for such movement. Furthermore, a large percentage (54 per cent) of automobile travel is for trips of five miles or less in length with a very small number of passengers per vehicle. Therefore, if alternatives to this type of travel could be found, or if the use of public transport could be more effectively encouraged for it, significant reductions in fuel consumption could be forthcoming. All types of travel (urban and inter-city passenger, goods movement, etc.) need to be thoroughly analysed to determine the potential for fuel economy and what changes are necessary to bring about such economy.

Another essential aspect of transportation is the fact that people and goods move or are moved in response to a demand that requires a change in location; only a small percentage of travel is for the purpose of travelling. As a result, in many cases in order to effect reductions in energy utilisation the demand for travel must be altered, either by restriction, by the availability of substitutes, or by both. The legal and institutional measures necessary to restrict demand are highly complex and require close scrutiny.

The continuing growth of suburban residential communities and, more recently, suburban commercial and industrial complexes, has put added stress upon public transport systems and increased the need for automobile travel in some cases. Control of urban and suburban development patterns to provide integrated communities combining residential, work and recreational activities, can help to alleviate the continuing suburbanisation process and therefore, affect fuel consumption within these areas. However, development requires a better understanding of the interactions between transportation and land-use as well as major legislative and institutional alterations, in order to ensure a more transport-economical arrangement as well as fulfilling other requirements.

It is also imperative to examine whether and to what extent certain measures might be counter-productive and what their impact upon personal mobility and the quality of life might be. For example, a move to smaller, lighter and less fuel-consuming automobiles increases crash-injury potential, particularly in traffic streams which include larger and heavier goods-moving vehicles. Again, restrictions such as fuel rationing or driving bans decrease personal mobility and freedom and adversely affect industries related to leisure pursuits.

To overcome the limitations inherent in policy development based upon a single transport technology or sector, it is necessary to devote research to the various systems of energy utilisation, within the transport sector. These systems include urban transport, inter-city transport, urban goods transport, and the like. It would be desirable to examine how it is possible to meet the demand for people and goods while reducing total energy consumption through a better combination of the various individual and public means of transport. Similarly, the need exists to examine the possibilities of reducing the demand for travel through improved communications and goods delivery techniques.

The same applies to the other systems of energy utilisation such as the industrial systems, which should be the subject of a comprehensive study covering the entire chain of processes from raw materials to finished products and their utilisation, and even their possible recycling, with a view to reducing the total amount of energy needed throughout the chain. Similar considerations apply to the various services involved in urban systems.

At this stage, while science and technology can certainly make a substantial contribution, as well as benefit from a certain amount of guidance, they can only form part of an integrated effort in which systems analysis and the social sciences play a fundamental part. The latter are also essential for studying the process of change from present to future systems, perhaps an even more difficult problem.

VI ENERGY SYSTEMS

The preceding chapters have dealt with the R & D problems involved in producing energy from various sources and in converting, transporting, storing and using it. However, it is important to bear in mind that the different sources and the various energy sectors are not independent, but may depend on, compete with, or complement one another. These relationships may be seen for example, in the connection between energy carriers and energy resources on the one hand and with the utilisation of energy on the other. There is obviously competition between the various energy resources or between the various energy carriers, but this is not incompatible with a certain degree of complementarity: e.g. the use of high temperature nuclear reactors for the gasification of coal and the use of methanol in fuel cells.

The energy field cannot therefore be regarded simply as a number of co-existing sources, technologies and applications, but as a veritable system. Therefore, if research is to serve a coherent energy policy, consideration must not be confined to the various energy technologies taken individually; energy systems will have to be treated as a whole, particularly where the links between energy, the economy, society and other natural resources(1) are concerned: the need for this has already been demonstrated in the preceding chapter with regard to the interaction between energy systems and utilisation systems, but it applies equally to all other aspects. Such systems will help in the formulation of energy policies and also contribute to the evolution of overall systems.

Economic problems obviously constitute one of the many aspects of the sector which must be studied from the standpoint of overall energy systems. An example is the problem of forecasting the supply of and demand for energy in its various forms. These forecasts have hitherto been made in a somewhat rudimentary fashion, often simply by extrapolating past trends, and have sometimes proved highly inaccurate, partly because not enough allowance was made for the relationships between the various sources and between the various carriers of energy. One objective of economic and econometric research should therefore be to produce more accurate forecasts through a better understanding of the supply and demand structure,

1) See Part I

price elasticity, the possibility and technical limits of substitutions, crossed elasticity coefficients and so on. Another example is the question of energy costs: decisions have usually been based on comparisons between the costs of individual technologies, whereas such comparisons ought to be made between the systems in which each technology is only one of the many parts.

In connection with the economic aspects, it is equally necessary to study all the effects which the growth and evolution of energy systems have on industrial systems. In particular, it is important to look into the likelihood of bottlenecks which - directly or indirectly - prevent or delay the introduction of new technologies. A particular problem here is that of the engineering firms which play a key role in the construction of many energy plants. Another point to be stressed in this connection is the capital importance of skilled manpower, especially engineers, for the development of energy systems; in view of the time that training takes, it is urgent to evaluate future requirements - both quantitative and qualitative - in this field and to consider ways and means of supplying them.

In order to better control the economics of energy systems as well as their interactions with other systems, the first essential is to improve our currently too limited understanding of the very structure of these systems, with special emphasis on two lines of research. First of all, it would be extremely useful to develop "energy accounting", i.e. a comprehensive survey of the flow of the various forms of energy, expressed in terms of physical quantities, from the level of primary energy resources up to that of the many different final uses. Secondly, the study of "energy geography" should be pursued, one of the primary objectives being to bring out the distribution of energy requirements and the other factors which affect the siting of energy production and transport facilities.

Energy geography is especially necessary as a means of studying the interaction between energy systems and those of the other natural resources - water and land. Reference has already been made to water resources in connection with electricity and the problem is very likely to be one of the major difficulties facing many countries in the coming decades. It is an increasingly important constraint on the siting of most energy plants (not only power stations). For this reason, it should be studied in close conjunction with land use, a problem which will be another of the principal constraints on the development of energy systems.

Energy geography and energy accounting need to be developed for another purpose - the study of the relationships between energy systems and the environment. Tackling pollution problems in relation to individual technologies is not enough, they must be considered in the broader framework of the pollution generated by all these technologies put together with particular reference to the problem of their siting. It is also important to devote far more R & D to the ecological and climatic effects of producing and consuming large amounts of energy. Two questions call for special attention: the ecological effects of increasing the amount of carbon dioxide in the atmosphere as a result of burning vast quantities of fossil fuels, and the effects of producing and using energy on the local, regional or even world climate.

There is also a need for more research into the safety aspects of energy systems, which are related to both environmental and social aspects of energy. These questions have come to the forefront as a result of public reaction to the development of nuclear power, but they need to be considered in a much broader context and merit appreciably more R & D, since up to now this has been largely confined to nuclear power questions.

Lastly there is the important question of the relationships between energy systems and the structure and life style of industrial society. Reference was made to this question in the previous chapter when discussing the use of energy and a further reminder is called for in the present context of the need to study all the various systems of energy usage, in relation to the psychological, sociological and ecological aspects of modern life.

In all research concerned with general energy systems, the techniques of systems analysis provide an extremely useful tool. In this connection encouragement should be given to work on methodology designed to improve these techniques and adapt them to the special case of energy systems. However, it should be emphasized that systems analysis must be the necessary complement to, not a substitute for, research in the various energy technologies or in the specific problems of the different scientific disciplines. Its special role is to provide a conceptual framework into which research results may be fitted, to identify the problems involved in the combined use of different technologies in energy systems and thus to signpost new priorities for research and development.

Research into energy systems should therefore be an important factor in framing an energy policy as well as scientific and technological policies in the energy field. Such research should also lead to a better understanding of the precise structure of existing energy systems and permit a more comprehensive evaluation of the characteristics, advantages and drawbacks of the various energy systems that may be proposed for the future. <u>From this point of view, research into energy systems ought therefore to prove extremely useful as an aid to decision-making, with regard to the framing of medium and long-term energy policies</u>.

Part III

ENERGY R & D POLICIES IN OECD MEMBER COUNTRIES

INTRODUCTION

Part III attempts to analyse the response of OECD governments to the new energy challenges, as far as they express themselves in science and technology policies.

This attempt, which is based mainly upon government replies to an OECD questionnaire, has certain limitations. Not all countries have replied(1), and those who did reply concentrate on government policies and partly omit industrial energy R & D. Some replies have in part been useless for the present survey, and even in the best cases statistical data have often been difficult to compare, as they were based on different years and definitions. To these difficulties, should be added the fact that energy R & D plans had not been finalised in all countries at the time the replies were written, that is between December 1973 and May 1974. In some cases, changes have occured or will occur, and some of the data presented here may no longer be completely up-to-date.

The tables appearing before Chapter 1 summarise and compare as well as possible the available data on Member countries R & D programmes, contained in Annex I, and R & D expenditures in the different energy sectors (excluding the environment). The analysis of Part III refers to a large degree to such data except for Chapter VI "Energy R & D and Developing Countries", which is not based on country replies.

As mentioned earlier, the statistical data provided by Member countries are based on different interpretation and definitions, due to the absence of a precise common typology of energy R & D expenditures. Tables 1 to 8 provide therefore figures which must be examined and compared with extreme caution. In many cases for instance the data supplied relate only to public expenditure whilst in others both public and private expenditure are included, thus showing an apparently great difference between one country and another. Another case in point is the difficulty of distinguishing

---

1) Replies have come from Australia, Austria, Belgium, Canada, Denmark, Finland, France, Germany, Greece, Ireland, Iceland, Japan, Luxembourg, Netherlands, Norway, Sweden, Switzerland, Turkey, United Kingdom, United States. No reply has been received from: Italy, Portugal, Spain and Yugoslavia.

the part of research from that of purely industrial activity, as for instance between research and exploration related to resources in hydrocarbons, or the difficulty of identifying expenditures related to energy conservation, which in many countries are included in other categories such as energy conversion, etc. Finally, for given years or sectors, the expenditures listed by some countries relate only to those particular research projects which could be identified at the time of the survey.

In spite of limitations, the country replies reveal a number of general and very fundamental facts. Few OECD countries seem to have had a general energy and energy R & D policy and none was prepared for the "energy crisis" which resulted from the oil-embargo in October 1973. In the past, energy and energy research policies of OECD countries have been largely sectorial, focussing, for example, on oil or coal or nuclear power. Overall co-ordination between these relatively independent policies or between the responsible agencies has been weak or non-existent.

As far as could be judged from the country replies, the old piece-meal approach to energy R & D had not yet been really overcome in early 1974. The intention to change short-sighted policies of the past has been clearly stated in several countries, and many steps towards a more global approach are visible, but a comprehensive, long-term national policy with clear targets cannot yet be found underneath the many different energy R & D programmes, at least not in the majority of the countries concerned.

ENERGY R & D EXPENDITURES

IN MEMBER COUNTRIES

EXPLANATORY NOTES

a) Tables 1 to 8 summarize energy R & D expenditures for 1972, 1973 and 1974 only, which are indicated in the case of most countries in Annex I, excluding expenditures concerning the Environment. Expenditures related to preceding or subsequent years have been listed in Annex I, together with the description of the relevant programmes, with the exception of Belgium whose expenditures for 1971 have been included in the tables as no other years were given.

b) The column "global expenditures" included in some tables, does not represent the total of the different columns for a given country, but indicates global figures related to several programmes within a given field for which no detailed breakdown has been made available.

c) In view of the inadequacy of the data concerning expenditures for one given year or field of activity, the figures appearing in the Column "Total" must be considered illustrative rather than definitive. The Secretariat has nevertheless deemed useful to include such totals in the tables, when they have appeared meaningful for a given year or sector, as they provide an indication of the order of magnitude of countries' expenditures.

d) Unless otherwise specified, R & D expenditures listed in the tables are government supported.

e) The conversion of national currencies into U.S. dollars has been made on the basis of "Wednesday averages" exchange rates of the New York Stock Exchange for 1972 and 1973. For 1974, figures have been converted at the same rate as 1973. (See Annex V)

f) In cases where countries have provided "annual" expenditures, the figures are followed by the mention "p.a." (per annum) and are given for the year 1974 only unless countries have specified the duration of the related programmes.

Table 1

ENERGY R & D EXPENDITURES IN MEMBER COUNTRIES

- 1972, 1973, 1974 -

SUMMARY TABLE

(All Sources)

In thousands US dollars

Country	Year	Fossil Fuels			Nuclear Energy	Other Energy Sources				Energy Carriers	Utilisation and Conservation of energy	Energy Sys-tems	Environ-ment	Global Expendi-tures	Total (See explanatory notes p.103)
		Hydro-carbons	Coal	Total		Solar Energy	Geothermal Energy	Other	Total						
Australia	1972-73		7,080	7,080	19,320										1972-73: 26,400 (a)
	1974	119	392	511	56	520			520		360		217		
Belgium	1971	102	1,158	1,260	19,360					1,980	880	150		19,964(b)	1971: 43,600 (c)
Canada	1972-73 1973-74	15,450	2,700	18,150	74,350	5,220			5,220	6,320	180		6,315		1973-74: 105,315
Denmark	1974				3,670	30			30	300	700				1974: 4,700
Finland	1972			34											
	1973			35						50					
	1974				1,480						150				1974: 1,630
France	1973	81,008	18,359	99,370	225,600	1,107	782	65	1,950	72,240	14,020		10,960		1973: 424,140 (d)
	1974	33,636	18,749	52,390	224,080	1,715	1,346	109	3,170	37,870	23,500		11,700		1974: 352,710
Germany	1973				215,220										
	1974	4,813	118,610	123,430	299,800	1,000	344	1,719	3,060	5,680	23,380	1,720			1974: 457,000 (e)
Iceland	1974								1,600						1974: 1,600
Ireland	1972		306	306	10					26					1972: 342
	1973		315	315	10					24					1973: 349
	1974		375	375	10					29					1974: 414
Japan	1972				125,950		300		300						
	1973	6,700		6,700	158,190		1,225	560	1,790	1,100	630				1973: 168,410
	1974	13,790	1,616	15,400	158,000	3,243	2,996	3,694	9,930	1,520	3,000				1974: 187,850 (f)
Netherlands	1973			500	77,000				3,300		2,600	400	900		1973: 84,700
	1974				29,190										
Norway	1973	4,040		4,040						5,460					1973: 9,500 (g)
	1974				5,080										1974: 5,080
Sweden	1973-74	105		105	9,360		100		100	2,460	430				1973-74: 12,460 (h)
Switzerland	1974	920		920	3,830	31		153	180	520	40		612		1974: 6,100 (i)
United Kingdom	1971-72				141,830										
	1972-73	2,900	14,200	17,100	167,000					17,100	6,000	2,400			1972-73: 209,600
	1973-74	4,000	17,895	21,900	164,000	770		240	1,010	19,400	15,700	2,400	4,150		1973-74: 228,660
United States	*FY 1973	18,700	85,100	103,800	481,300	4,000	4,400	900	9,300	11,000	21,200	7,200	38,400		FY 1973: 672,200
	*FY 1974	19,100	164,400	183,500	633,400	13,800	10,900	11,500	36,200	23,800	42,200	17,300	65,500		FY 1974: 1,001,900

* Fiscal year.

NOTES:

a) Of this total, the amount of US$4,091,000 relative to expenditures on coal research is supported by industry.

b) Industry R & D expenditures for non-nuclear energy.

c) Of which US$19,964,000 supported by industry for research on non-nuclear energy.

d) Of which a total of US$98,464,000 is supported by industry for all energy sectors except the utilisation and conservation of energy.

e) Only two-thirds of the expenditures related to Fossil Fuels, included in this total, are supported by government.

f) Does not include funds related to the "Sunshine" project.

g) Of which US$4,135,000 are supported by industry.

h) For industry contribution to R & D expenditures see Table 5(note h).

i) For industry contribution to R & D expenditures see Table 2, 3, 4 and 8.

Table 2

ENERGY R & D EXPENDITURES IN OECD MEMBER COUNTRIES 1972, 1973, 1974

FOSSIL FUELS

By Type of Fuel and Programme

In thousands of U.S. dollars

Country	Year	1 Hydrocarbons				2 Coal					3	4
		Resources Assessment	Oil	Natural Gas	Shale Tar Sands Heavy Oil	Resources Assessment	Coal Mining	Improved Combustion	Coal Conversion	Other	Global Expenditures	Total (See explanatory notes p. 103)
Australia	1972-73						7,080(a)					1972-73: 7,080
	1974	119p.a.				77p.a.	42(b)	273				
Belgium	1971		102			1,158						1971: 1,260(c)
Canada	1973-74	1,463	13,993			156	1,348		117	1,079(d)		1973-74: 18,150
Finland	1972 1973									34 35		1972: 34 1973: 35
France	1973 1974	21,419(e) 10,373	59,589(f) 22,807		456	18,359 18,749						1973: 99,370(g) 1974: 52,390
Germany	1974		4,813				35,060	9,280	65,670	8,600		1974: 123,430(h)
Ireland	1972 1973 1974						 24p.a.			306 315 351		1972: 306 1973: 315 1974: 375
Japan	1973 1974		5,960(i) 13,790						 1,616		740(j)	1973: 6,700 1974: 15,400
Netherlands	1973										500	1973: 500
Norway	1973	4,040(k)										1973: 4,040
Sweden	1973-74		83	22								1973-74: 105
Switzerland	1974		920p.a.(l)									1974: 920
United Kingdom	1972-73 1973-74		2,900 4,000			240 240	9,300 11,000	1,030 1,125	690 2,170	2,900 3,360		1972-73: 17,100 1973-74: 21,900
United States	FY* 1973 FY 1974	4,500 5,000	11,000 11,800		3,200 2,300	1,000 1,200	29,900 35,800	1,500 15,900	48,100 99,800	4,600 11,700		FY1973: 103,800 FY1974: 183,500

* Fiscal year

NOTES:

a) Expenditures supported by government/university (US$2,993,000) and industry (US$4,091,000).

b) Part of a 3-year programme totalling US$126,000.

c) Public expenditures. Total industry R & D expenditures for non-nuclear energy in 1971 amounted to US$19,964,000. No detailed breakdown is available for such expenditures showing the part devoted to Fossil Fuels.

d) Includes transportation, environmental management and conservation.

e) Includes expenditures by government (US$8,594,000) and nationalised and/or private industry (US$12,325,000). 1974 figures refer to public expenditures only.

f) Includes expenditures by government (US$20,854,000) and industry (US$38,735,000). 1974 figures refer to public expenditures only.

g) of which $51,560,000 are supported by nationalised and/or private industry.

h) 2/3 of these expenditures are government supported.

i) Part of a 5-year programme (FY1970-FY1974) amounting to a total of 5 billion yen.

j) Expenditures relate to storage of oil on the sea bed. The programme starts from FY1973 but as no duration has been indicated it is not sure whether the expenditures refer to that year only.

k) Covers exploration of oil and gas.

l) Expenditures supported by industry with the participation of the government.

Tabl

ENERGY R & D EXPENDITURE

NUCLEA

Country	Year	1. Uranium, Resources (Exploration, Assessment Exploitation)	2. Fuel Cycle		3. HWR	4. LWR	5. HTGR	6. Fast Breeders (a)
			Uranium Enrichment	Other				
Australia	1972-73 1974	56p.a.						
Belgium	1971 1973 1974					2,806	306	5,134 12,220 12,220
Canada	1973-74	869			64,877(c)			
Denmark	1974							
Finland	1974	373p.a.				149p.a.		
France	1973 1974			9,852 9,964		50,583 45,874	16,840 19,031	96,652 100,645
Germany	1973 1974	9,626	39,193	24,067		10,314 7,907	77,011 68,416	79,762 89,732
Ireland	1972 1973 1974							
Japan	1972 1973 1974	478 665 1,126	6,528 20,310 36,394	5,693 4,680 3,934	864 1,470 1,293		1,363	65,288 65,210 44,996
Netherlands	1973 1974							29,190(h)
Norway	1974			496p.a.		3,890p.a.(i)		
Sweden	1973-74	1,566				4,629	625	
Switzerland	1974	31p.a.					1,530p.a.(j)	122p.a.
United Kingdom(l)	1971-72 1972-73 1973-74	 240 240	 5,520 10,100			10,256	12,507 14,400 12,360	75,541 80,160 79,740
United States	*FY 1973 *FY 1974	2,800 3,400	50,300 57,500				7,300 13,800	259,300(m) 361,300(n)

* Fiscal Year

NOTES:

a) Except when it is otherwise specified, expenditures listed in this column relate to LMFBR.

b) Expenditures for research carried out in industry almost totally supported by government. The detailed amount for the different nuclear programmes is not available as far as these expenditures are concerned.

c) R & D expenditures on HWR(CANDU) and advanced fuel cycles.

d) Partly financed by EURATOM.

e) Industry expenditures excluding research carried out by industry but supported by government.

f) Co-operative programme Ireland (60 per cent); UK-UKAEA (30 per cent); France-CEA (10 per cent).

g) Part of a ten-year programme amounting to more than 200 billion Yen.

N OECD COUNTRIES 1972, 1973, 1974
ERGY

In thousands of U.S. Dollars

7. Other Reactors	8. Safety	9. Radio Active Waste Management	10. Underlying Research	11. Thermonuclear Fusion	12. Nuclear Ship Propulsion	13. Global Expenditures	14. Total (See explanatory note p. 103
						19,320	1972-73: 19.320
	330		1,895	369		8,518(b)	1971: 19,360
	1,081		7,518				1973-74: 74,350
	3,000p.a.	70p.a.		600p.a.			1974: 3,670
	461		499				1974: 1,480
,796 ,668	25,433 29,578			12,565(d) 14,995	239 326	8,646(e)	1973: 225,600 1974: 224,080
	18,221 24,756	5,157p.a.		20,284 23,378	9,626 7,564		1973: 215,220 1974: 299,800
				10) 10)(f) 10)			1972: 10 1973: 10 1974: 10
,078(g) ,367 ,996	1,175 5,256 14,537	204		2,158 2,317 3,680	5,690 4,918 5,553		1972: 125,950 1973: 158,190 1974: 158,000
						77,000 29,190	1973: 77,000
	413	280					1974: 5,080
	1,339	67	832	300			1973-74: 9,360
765p.a.		310p.a.(k)		1,070p.a.			1974: 3,830
,257 ,800 ,160	7,200 7,200	3,600 3,840	19,261 17,520 17,280	11,006 9,600 7,200			1971-72: 141,830 1972-73: 167,000 1973-74: 164,000
,500 ,000	38,800 48,600	3,600 6,200	14,900 10,700	74,800 101,100	1,800p.a.		*FY 1973: 481,300 *FY 1974: 633,400

Multinational programme: Belgium, Germany, Netherlands (15 per cent).

Of this amount $3,300,000 represents work at Halden operated as an OECD project. Includes research on: fuel element development and testing; fuel management and core physics; computer control of power reactors; safety and radio-active waste. Financed by Norwegian Government (one third) and by signatories of OECD Halden project (two-thirds).

Bilateral programme (Germany-Switzerland)

Financing by a group of utilities with government participation.

Figures given for different programmes do not include: UKAEA International grants amounting to:
$3,008,000 - 1971-72
$3,389,000 - 1972-73

Includes $5,600,000 for GCFBR and MSFBR

Includes $4,000,000 for GCFBR and MSFBR

Table 4

ENERGY R & D EXPENDITURES IN OECD MEMBER COUNTRIES

1972, 1973, 1974

OTHER ENERGY SOURCES
By Source

In thousands U.S. dollars

Country	Year	1 Solar Energy	2 Geothermal Energy	3 Urban and Industrial Waste	4 Wind, Tides, Ocean Currents	5 Other (non-specified)	6 Global Expenditures	7 Total (See explanatory notes p. 103)
Australia	1974	520						1974: 520
Canada	1972-73						5,220(a)	1972-73: 5,220
Denmark	1974	30 p.a.						1974: 30
France	1973	1,107	782		65			1973: 1,950
	1974	1,715	1,346		109			1974: 3,170
Germany	1974	1,000	344 p.a.	1,719 p.a.				1974: 3,060
Iceland	1974		1,600 p.a.(b)					1974: 1,600
Japan	1972		300					1972: 300
	1973		1,225	560				1973: 1,790
	1974	3,243	2,996	1,494		2,200		1974: 9,930
Netherlands	1973						3,300	1973: 3,300
Sweden	1973-74						100	1973-74: 100
Switzerland	1974	31 p.a. (c)		153 p.a.				1974: 180
United Kingdom	1973-74	770		240				1973-74: 1,010
United States	*FY 1973	4,000	4,400	(d)	(e)	900		FY 1973: 9,300
	*FY 1974	13,800	10,900			11,500		FY 1974: 36,200

* Fiscal year.

NOTES:

a) This amount represents total federal 1972-73 R & D expenditures for research in the following fields:
Solar energy; chemical batteries; other storage techniques; conversion (biomass and waste); geothermal energy; hydropower; tidal energy; wind generators.

b) This amount covers research programmes in the field of hydropower and geothermal energy. No detailed breakdown is available.

c) Financing is covered up to 70 per cent by public funds and 30 per cent by private resources.

d) Included in solar/environmental total.

e) Included in solar total.

Table 5

ENERGY R & D EXPENDITURES IN OECD MEMBER COUNTRIES
1972, 1973, 1974

ENERGY CARRIERS

In thousands U.S. dollars

Country	Year	1. Electricity			2	3	4	5
		a) Generation	b) Storage	c) Transmission	Heat and District Heating	Other energy carriers	Global Expenditures	Total (See explanatory notes p. 103)
Belgium	1971						1,980(a)	1971: 1,980
Canada	1973-74	2,018(b)		4,301				1973-74: 6,320
Denmark	1974				300 p.a.			1974: 300
Finland	1973				50			1973: 50
France	1973 1974						72,240(c) 37,870(d)	1973: 72,240(c) 1974: 37,870(d)
Germany	1974	516 p.a.	1,547 p.a.	2,235 p.a.	1,380 p.a.			1974: 5,680
Iceland	1974	1,600 p.a.(e)						1974: 1,600
Ireland	1972 1973 1974	13 12 17		13 12 12				1972: 26 1973: 24 1974: 29
Japan	*FY 1973 *FY 1974	1,100(f) 1,520						FY 1973: 1,100 FY 1974: 1,520
Norway	1973	1,323		4,135(g)				1973: 5,460
Sweden	*FY 1973-74	434	859	1,000(h)	162(i)			FY 1973-74: 2,460
Switzerland	1974	460 p.a.		31 p.a.	24 p.a.(j)			1974: 520
United Kingdom	1972-73 1973-74	11,280 13,500	607 533	4,968 4,632	240 720	48		1972-73: 17,100 1973-74: 19,400
United States	*FY 1973 *FY 1974	6,500 15,900	1,600 2,900	2,900 5,000	(k)			FY 1973: 11,000 FY 1974: 23,800

* Fiscal year

NOTES:

a) Public expenditures. Total non-nuclear expenditures by industry, in 1971, amount to U.S. $19,964,000. No detailed breakdown is available of such expenditures, showing the part devoted to energy carriers.

b) Excludes generation from nuclear sources.

c) Includes expenditures by Government (U.S. $33,983,000) and industry (U.S. $38,258,000).

d) Government expenditures.

e) Includes research on geothermal energy.

f) Part of a ten year programme (1966-75) amounting to a total 6.4 billion Yens.

g) Industry supported programme.

h) Rough expenditures supported by government and industry.

i) Of this amount, U.S. $54,000 relate to an industry supported programme.

j) Part of a five year programme amounting to a total of U.S. $122,000.

k) Included in residential and commercial column of Table 6.

Table 6

ENERGY R & D EXPENDITURES IN OECD MEMBER COUNTRIES

1972, 1973, 1974

UTILISATION AND CONSERVATION OF ENERGY
By Sector

In thousands U.S. dollars

Country	Year	1 Residential and Commercial	2 Industry and Agriculture	3 Transportation	4 Global Expenditure	5 Total (See explanatory notes p. 103)
Australia	1974	360 p.a.				1974: 360
Belgium	1971				880	1971: 880
Canada	1973-74				180	1973-74: 180
Denmark	1974		700 p.a.			1974: 700
Finland	1974	150(a)				1974: 150
France	1973 1974	4,102 5,816	7,335 14,257	2,583 3,429		1973: 14,020 1974: 23,500
Germany	1974			20,630 p.a.	2,750(b)	1974: 23,380
Japan	*FY 1973 *FY 1974		630(c) 3,000			FY 1973: 630 FY 1974: 3,000
Netherlands	1973				2,600	1973: 2,600
Sweden	*FY 1973-74	317(d)	110(e)			FY 1973 -74: 430
Switzerland	1974	40 p.a.				1974: 40
United Kingdom	1972-73 1973-74	 7,850	6,000 5,900	 1,920		1972-73: 6,000 1973-74: 15,700
United States	*FY 1973 *FY 1974	 15,000		21,200 27,200		FY 1973: 21,200 FY 1974: 42,200

* Fiscal year.

NOTES:

a) Part of a 4 year programme (1974-77) amounting to a total of US$598,000.
b) Includes research on electricity generation and transmission and on district heating.
c) Part of a 6 year programme amounting to $4.5 million.
d) Includes US$43,000 for district heating.
e) Of this amount US$ 16,000 are financed by the Royal Swedish Academy of Sciences.

Table 7

ENERGY R & D EXPENDITURES IN OECD MEMBER COUNTRIES
1972, 1973, 1974

ENERGY SYSTEMS

in thousands US dollars

Country	Year	Expenditures	Remarks
Belgium	1971	150	
Germany	1974	1,720	
Netherlands	1973	400	
United Kingdom	1972-73 1973-74	2,400 2,400	
United States	*FY 1973 FY 1974	7,200 17,300	

* Fiscal year

Table 8

ENERGY R & D EXPENDTTURES IN OECD MEMBER COUNTRIES
1972, 1973, 1974

ENVIRONMENT

in thousands US dollars

Country	Year	Expenditures	Remarks
Australia	1974	217 p.a.	
Canada	*FY 1973-74	6,315	
France	1973 1974	10,960 11,700	
Germany			Expenditures for R & D on environmental effects of energy transformations are included in Table 3, column 8 (safety).
Netherlands	1973	900	
Switzerland	1974	612 p.a.	Government supported programme with participation of industry
United Kingdom	1972-73 1973-74	 4,150	
United States	*FY 1973 *FY 1974	38,400 65,500	

* Fiscal year

ENERGY R & D ACTIVITIES IN MEMBER COUNTRIES

The following lists have been drawn up on the basis of the material contained in Annex I as well as of the data available from the survey of the Energy Co-ordinating Group(1) created by the Washington Conference in February 1974.

1) Report of the Energy Co-ordinating Group - Ad Hoc Group "International Co-operation on Energy Research and Development" ECG/ERD/36 final, 6th June 1974.

Table 9

ENERGY R & D ACTIVITIES IN MEMBER COUNTRIES

FOSSIL FUELS

Field of activity \ Country	Australia	Austria	Belgium	Canada	Denmark	Finland	France	Germany	Greece	Iceland	Ireland	Italy	Japan	Netherlands	Norway	Sweden	Switzerland	Turkey	United Kingdom	United States	Existing International R & D Co-operation
Resource Assessment and prospection																					
1. Advanced Prospection techniques			X	X	X		X	X					X						X	X	NASA; EURATOM; International Union of Geological Sciences; US - Japan; US-Multinational
2. Oil and Gas Prospection	X	X		X	X		X	X	X		X	X	X	X	X		X	X	X	X	
3. Coal Prospection	X			X														X	X	X	
Hydrocarbons																					
4. Oil: Drilling, Production (R & D)							X	X					X	X	X				X	X	
5. Oil: Transportation, Pipelines				X								X		X	X				X	X	
6. Natural Gas: Production, Conversion							X													X	
7. Natural Gas: Transportation - Gas or LNG -				X			X							X					X	X	
8. Oil and Gas: Storage (underground, undersea, other)	X				X		X	X				X	X	X		X	X		X	X	European Gas Research Group
9. Tar Sands: Prospection, Production				X									X							X	
10. Shale oil: Prospection, Production				X			X	X										X		X	
Coal																					
11. Coal: Mining	X		X	X			X	X		X									X	X	
12. Coal: Improved Combustion, Pollution Abatement	X			X			X	X						X					X	X	ECSC; US-Germany; US-UK; US-Japan; US-EC (items 11-14)
13. Coal: Gasification				X				X					X	X					X	X	
14. Coal: Liquefaction								X					X						X	X	
15. Peat: Prospection, Production, Combustion						X			X		X		X								Ireland-Finland, - Germany Sweden
16. Coal: Other problems (coking, Transportation)	X		X	X			X	X					X	X					X	X	ECSC

Table 10

NUCLEAR ENERGY

Field of activity \ Country	Australia	Austria	Belgium	Canada	Denmark	Finland	France	Germany	Ireland	Italy	Japan	Netherlands	Norway	Sweden	Switzerland	U.K.	U.S.	Existing International Co-operation
1. Uranium Resources																		
- Exploration Assessment and Exploitation	x	x		x	x	x	x	x	x					x	x	x	x	
2. Fuel Cycle																		
- Uranium Enrichment	x		x				x	x			x	x		x		x	x	EURODIF
- Fuel Processing			x	x	x			x		x	x	x				x	x	IAEA-NEA
- Fuel Reprocessing			x				x	x		x	x		x		x	x	x	Germany-France-U.K.; EUROCHEMIC
3. Heavy Water Reactors				x				x		x	x		x			x		Canada-Japan; - Australia, - U.S. - , Sweden
4. Light Water Reactors			x		x	x	x	x		x	x	x	x	x			x	U.S.-EURATOM; U.S.-IAEA; OECD(HALDEN); 30 bilateral agreements with U.S.
5. High Temperature Gas Cooled Reactors			x				x	x			x	x	x		x	x	x	U.S.-Germany; U.S.-France; NEA (DRAGON)
6. Fast Breeders:																		
- LMFBR			x				x	x		x	x	x	x	x		x	x	Belgium-Netherlands-Germany; France-Germany-Italy; U.K.-U.S.-EURATOM; U.S.-Japan; U.S.-Netherlands
- Other			x					x		x	x	x	x			x	x	NEA; EURATOM
7. Nuclear Safety	x		x	x	x	x	x	x	x	x	x	x	x	x		x	x	EURATOM; NEA; MARVIKEN(U.S.Scandinavia-Japan-Germany); U.S.-Japan; U.S.-Belgium; U.S.-Germany
8. Radioactive Waste Management	x	x	x	x	x		x	x		x	x	x	x	x	x	x	x	IAEA; EURATOM; NEA; U.S.-NEA; EUROCHEMIC; U.S.-Japan
9. Thermonuclear Fusion			x	x	x		x	x	x	x	x	x	x	x	x	x	x	EURATOM; U.S.-U.K; 6 bilateral agreements with U.S.A.
10. Nuclear Ship Propulsion							x	x		x	x		x			x	x	IMCO; NEA; U.S.-Germany

Table 11

OTHER ENERGY SOURCES

Field of activity \ Country	Australia	Austria	Belgium	Canada	Denmark	Finland	France	Germany	Iceland	Ireland	Italy	Japan	Netherlands	Sweden	Switzerland	U.K.	U.S.	Existing International Co-operation
1. Solar Energy																		
- Heating and cooling of buildings and/or central power station and/or photovoltaic conversion	X		X	X	X		X	X		X	X	X	X	X	X	X	X	NATO (CCMS); EC-EURATOM; US - New Zealand; US - Australia; UNESCO; Ireland - UK; US - Japan; US - France
2. Geothermal Energy		X		X			X	X	X		X	X				X	X	NATO (CCMS); US - Iceland; US - New Zealand; NATO; US - Japan
3. Organic matter and waste																		
- Bioconversion of organic material and waste				X				X				X	X			X	X	
- Municipal and industrial waste				X	X			X				X	X	X	X	X	X	
4. Wind, Tides and Ocean thermal gradients																		
- Wind				X	X			X			X	X	X	X		X	X	
- Tides and Waves				X				X				X				X	X	
- Ocean thermal gradients						X						X					X	NATO

Table 12

ENERGY CARRIERS

Field of Activity \ Country	Austria	Belgium	Canada	Denmark	Finland	France	Germany	Iceland	Ireland	Italy	Japan	Netherlands	Norway	Sweden	Switzerland	U.K.	U.S.	Existing International Co-operation
Electricity Generation																		
1. Hydropower								X					X		X			
2. Improvement of power plants						X			X						X	X		
3. Reduction of thermal pollution						X			X		X			X	X	X	X	
4. Turbines and combined cycles		X		X		X	X				X	X	X	X		X	X	COST
5. MHD		X					X			X	X	X		X	X		X	IAEA; NEA; US-Germany, US-Japan; IAEA-NEA
6. Fuel Cells						X	X				X	X					X	US-UK; US-Japan
Electricity Storage																		
7. Batteries, hydraulic storage			X			X	X				X	X		X		X	X	
8. Flywheels							X					X					X	
Electricity Transmission																		
9. Underground cables		X	X				X						X			X	X	
10. Superconducting technologies		X	X			X	X			X	X	X	X		X	X	X	EC; US-Japan
11. High voltage lines	X	X	X				X		X	X	X	X	X	X		X	X	
Heat and District Heating																		
12. Dual purpose plant				X		X	X				X					X	X	IAEA; several bilateral programmes with US.
13. Heat pipes							X				X	X					X	
14. District heating				X	X		X					X	X	X	X	X	X	
Other Energy Carriers																		
15. Hydrogen						X	X				X	X			X	X	X	EURATOM; US-Japan
16. Methanol				X			X					X					X	

Table 13

a) - UTILISATION AND CONSERVATION OF ENERGY

b) - ENERGY SYSTEMS

Field of activity \ Country	Australia	Austria	Belgium	Canada	Denmark	Finland	France	Germany	Italy	Japan	Netherlands	Norway	Sweden	Switzerland	U.K.	U.S.	Existing International Co-operation
a) - Energy Utilisation																	
1. Residential and Commercial Sector	X	X	X		X	X	X	X	X		X	X	X	X	X	X	
2. Industry and Agriculture			X	X	X		X		X	X	X	X			X	X	IIASA; OECD (Road Research) NATO (CCMS); U.S. - France; U.S. Japan.
3. Transportation			X		X		X	X	X	X	X	X			X	X	
b) - Energy Systems																	
1. Overall energy systems analysis and general studies			X	X	X		X	X	X	X	X	X	X		X	X	IIASA; NEA; UNEP; WHO; OECD; IAEA; EC

I TOWARDS AN OVERALL STRATEGY

A. PROGRAMMES AND FINANCING

The direct consequence of the oil crisis has been an additional burden to the economy of oil importing countries which has in some cases further aggravated an existing balance-of-payments deficit. National economies are now facing, in addition, another indirect consequence of the crisis, i.e. the increase of energy R & D activities in an attempt to reach self sufficiency or at least to ease their dependence in energy supplies.

This is the first general trend which emerges from the observation of national programmes and budgets on energy R & D. In fact, not only have additional amounts been devoted to the development of the potential of conventional sources such as coal and oil through new technologies - coal gasification and liquefaction - or to the exploration of additional reserves, such as off-shore oil, but nuclear research has increased considerably, whilst an entirely new set of other sources have gained new financial support. Energy promises thus to be for the next decade and at least until alternative sources for oil are developed, one of the major objectives in R & D expenditures, both public and private which will call for a change in R & D priorities.

Obviously resources allocated to energy R & D should primarily be established on the basis of the energy objectives set by each individual country, hence of their energy policy. However, other factors related to science policy, as well as to economic, industrial and environmental policies, will also play a key role in determining the overall level of national energy R & D budgets.

Whatever the weight given to different factors to set a desirable level of R & D, what is essential is to ensure that the present sudden increase in energy R & D be not followed in the next few years by an equally sudden curtailment of expenditures.

The discontinuity of the research effort is more likely to occur in this field as the present and planned programmes are partly the result of the crisis. It is possible that once the more spectacular effects of the crisis disappear, programmes may be slowed down and reduced considerably even if the crisis continues to be real for many years. The risk of such a "stop and go" policy would then be an ill balanced research policy sacrificing long-term objectives to short-term needs.

1. The imperatives of a balanced R & D

R & D will play a key role in solving the energy problem but is not an isolated element of the energy issue. It is essential then that its programmes and budgets be well balanced so as to take into account the various components on which it will have repercussions. In quantitative terms this means that the distribution of resources should be directed towards keeping as wide a number of energy options as possible. In qualitative terms, the purpose of this balance is to avoid the development of a massive technology without creating the proper human and environmental conditions for its acceptance. If properly balanced, then R & D can help to avoid distortions and bottlenecks at different levels.

At the policy level, research aimed at energy production should no longer be limited to one primary source. Efforts should be made to develop alternative primary sources with a view to a fuel diversification policy. Such diversification can in the best case bring about complete independence as to supplies and in any case, reduce dependence from a single supplier and from a unique form of energy.

Equally at the policy level, the finite nature of certain primary sources of energy and the inability of countries to meet the needs of an exponential growth at past rates and under acceptable economic and environmental conditions, requires that research should aim not only at satisfying the energy demand but also at curbing it. There has been up to now no long-term view of the impact on the economy of energy conservation. This being a concept directly generated by the crisis, R & D plans in this field may be limited to short-term innovations and improvements to fight the crisis. Efforts should be made so that R & D on energy conservation be conceived in the future as part of a continuous long-term plan.

At the level of the energy cycle, a coherent energy development requires that a proportionate share of research be undertaken at the different stages of the energy cycle (production, conversion, storage and transmission) if bottlenecks of a technical nature are to be avoided.

At the level of energy systems. For the integration of energy in a comprehensive energy system, hardware development must be coupled to software research. Software research is needed not only on the technology side, i.e. to identify through system analysis the impact, cost and benefits of different technologies at the economic and environmental level, but also on the social side to examine the relation between energy and society and the problems related to this issue.

2. Overall review of energy R & D

The programmes and expenditures devoted by Member countries to energy R & D presented in this section, as well as in Annex I, are mainly drawn from information submitted before May 1974. For this reason and even taking into account the amendments transmitted later on, they do not in many cases represent the definitive plans of Member countries concerning energy R & D and may already have changed by the time this report is published.

Owing to marked differences in the energy situation of Member countries as regards both their natural resources and their R & D level, there is no ideal classification which would allow of meaningful generalisations and conclusions as to the content of energy R & D. However, for the purpose of this review, the OECD countries have been grouped according to the availability of natural resources, this being in most of them a major determining factor for the formulation of R & D programmes:

i) countries with large and diversified energy resources: Australia, Canada, Norway, United Kingdom, United States;

ii) countries with one predominant and relatively abundant energy resource: Iceland, Netherlands, New Zealand;

iii) other countries: Austria, Belgium, Denmark, Finland, France, Germany, Greece, Ireland, Italy, Japan, Luxembourg, Portugal, Spain, Sweden, Switzerland, Turkey, Yugoslavia.

a) Energy production: Choice and diversification of energy primary sources

Obviously the problem of choice and diversification of primary sources of energy is quite different for countries having considerable and varied natural resources and for those who have not, although the availability of resources has not necessarily been in the past a condition for a strategy of diversification. Indeed, very few countries have natural resources coupled with large national resources in terms of GNP and an adequate supply of manpower to be able to support research in the whole range of primary sources. The combination of these three elements may be found with different degrees of importance in three countries belonging to Group I, the United States, Canada and the United Kingdom.

Of the countries in Group I, and of all the countries covered, only the United States supports a systematic research effort on all energy sources. It is true that such diversification is possible because of the availability of natural resources and a high R & D expenditure capacity, but this represents nevertheless a new policy trend in a country - not the only one however - which has in the past disregarded the development of national energy resources in favour of increasing oil imports.

The total proposed federal expenditure on energy R & D for 1974 in the United States amounts to $1,000 million, of which 63 per cent for nuclear energy, 16 per cent for coal, 4 per cent for new energy sources, 2 per cent for hydrocarbons and the remaining 15 per cent for energy carriers, utilisation, systems and the environment. The energy R & D budget shows an increase of 49 per cent over 1973. However, the increase is comparatively much higher (400 per cent) for the development of new energy sources - mainly solar and geothermal energy - than for coal (100 per cent) or nuclear energy (35 per cent). Government support is slightly increasing for hydrocarbons, R & D in this field being predominantly financed by private firms. It should be noted that United States R & D expenditures on coal and new energy sources are equivalent to, if not higher than, the research effort made by all OECD countries together in these two fields.

The United States is also one of the few countries which is developing energy sources according to a long-term plan. The proposed "Project Independence" plans R & D for 5 years beginning in Fiscal Year 1975 and for that year the proposed increase in total energy R & D is 81 per cent over 1974.

Canada, with a smaller R & D potential, is engaged at present in three important research programmes. The first and most significant one in size (approximately 70 per cent of the entire R & D budget) relates to the development of an independent nuclear technology with the CANDU reactor. The second covers off-shore oil production systems and the third relates to tar sands production methods. Federal support for tar sands extraction is rather small (1 per cent of the total energy budget) but the provincial government of Alberta supports R & D on in-situ production of deep tar sands at the rate of 100 million Canadian dollars for a period of 5 years.

Canada also supports, in addition to a small programme on coal, a programme on new energy sources (5 per cent of the total energy R & D) covering the whole range of non-conventional sources other than nuclear. In absolute figures, the Canadian effort in this field is the highest one after the United States and Japan. With such a comprehensive programme and its large resources in uranium, oil, gas and tar sands (the largest deposits in the world), Canada is one of the few countries heading for self sufficiency.

The United Kingdom may be included in Group I countries particularly in view of the large reserves of oil and gas recently discovered in the North Sea which add to the substantial coal resources available. Expenditure on coal research is presently under review. In recent years nuclear research in the United Kingdom has concentrated on the fast reactor. With the recent decision on choice of thermal reactor for the United Kingdom programmes a changing pattern of expenditure is expected.

Australia, with relatively large resources in uranium, coal and oil makes a substantial effort particularly for coal In percentage of its total energy R & D this represents 27 per cent, a higher rate than Germany (26 per cent) or the United States (16 per cent), but in absolute figures the research effort on coal made by Australia is not comparable to that of those two countries. The largest expenditures (70 per cent) relate to nuclear energy and bear mainly on the development of some aspects of nuclear technology. A very modest effort, less than 2 per cent, is also made on new sources, namely on solar energy.

Last in this Group in respect of the extent of research programme undertaken, Norway makes nevertheless a considerable effort in the exploration of oil and gas following the important discoveries of the North Sea. R & D in the field of hydrocarbons is almost of the same magnitude as for nuclear energy ($4 million against $5.0 million for nuclear energy). Norway takes second place among Scandinavian countries after Sweden, for nuclear expenditures and can only support a modest share of an international programme. The extent of the resources in the North Sea cannot yet be accurately predicted but it is expected that Norway will be in a position to export large quantities of oil.

Countries included in Group II, having one predominant and relatively abundant resource, are engaged mainly in the development of that energy source. Research on alternative sources bears mainly on nuclear energy but the medium or small R & D potential of these countries does not allow the development of major independent programmes.

The Netherlands, with its large gas resources carries out an important privately funded programme in the field of hydrocarbons. The R & D programme for nuclear energy is in absolute figures the most extensive one of the countries in this Group. Programmes on other non-conventional sources have been announced but the related expenditures are not known. A national programme on energy research is being prepared.

Among the countries which are included in Group II, Iceland, has the smallest R & D potential and supply of scientific manpower as compared to other OECD countries. Last in this Group for the size of the research effort, this country concentrates all of its R & D potential on the exploitation of hydropower and geothermal energy. A total of $1.6 million is devoted to R & D in these two sources, but geothermal research has been increasing faster in recent years. Iceland is the only country not involved in any R & D programme in nuclear energy.

The third Group of countries with limited resources includes countries whose expenditures for the whole range of R & D activities and whose level of scientific manpower are either large: France, Germany, Japan; medium: Belgium, Sweden, Switzerland; or small: Austria, Denmark, Finland, Greece, Ireland, Luxembourg and Turkey.

France supports a large independent programme in the nuclear field and devotes considerable R & D funds to hydrocarbons ($81 million for 1973). Taking into account the fact that France has very limited hydrocarbon resources, such an effort is the result of a deliberate governmental policy at the national as well as at the international level. Expenditures for research on other sources such as solar and geothermal energy amount to 1 per cent of the total energy R & D budget in 1974.

Germany is one of the countries whose level of expenditures for energy R & D is very high, particularly taking into account that R & D expenditures mentioned in this report do not include industry supported research programmes. In 1974, Germany allocated 67 per cent of its total energy R & D budget to nuclear energy, which allows for the development of a largely independent programme, and 26 per cent to coal whose reserves are the largest ones after the United States in the OECD zone. In addition, a long-term 4-year programme is planned in almost all energy sectors and it is foreseen that research in nuclear energy and coal will increase by 50 per cent by 1976 and more than four times so far as new technologies for the exploitation of natural gas and oil resources are concerned.

Japan, an R & D intensive country with large supplies of manpower, makes a special diversification effort. Its independent programme on nuclear research is growing gradually and represented in 1974 about 85 per cent of the total budget. The attention given to coal development is increasing at an equal scale. In addition, through its Sunshine project a special effort is being planned on other non-conventional energy sources. Although no definite figures are available, the proposed research effort on solar and geothermal energy and waste processing may amount to about 8 per cent of the total energy R & D budget.

For Greece, the recent and potentially most important discovery of oil and gas in the Aegean Sea might constitute the major energy supply, although there is no indication as yet to the total amount of the deposits.

For the time being, R & D in Greece relates mainly to exploration of energy sources such as lignite, peat and geothermal energy besides oil and gas. The Greek Public Power Corporation alone will devote $6.5 million a year to R & D on energy with main emphasis on development, design and testing.

With the exception of Ireland which devoted in 1974 only 2 per cent of its energy budget to nuclear energy (90 per cent being allocated to research on peat) and of Luxembourg which makes no research effort in this field, most of the other countries in this Group devote a very high share of their budget to nuclear energy: Belgium, 44 per cent; Switzerland, 55 per cent; Sweden, 75 per cent; Denmark, 78 per cent; Finland, 86 per cent. Sweden is however the only country where the predominant nuclear research is linked to an advanced nuclear technology and a powerful national nuclear industry. The other countries carry out nuclear research either as part of a multinational effort (Switzerland, Belgium) or develop only some aspects of nuclear technology (Denmark, Finland).

Austria and Turkey have not provided sufficiently adequate figures to allow of meaningful comparison of the funds devoted to different energy primary sources.

b) Production versus conservation

Most countries appear to be aware of the need of including conservation programmes in the formulation of an energy policy on ethical and rational grounds. Research in this field still appears to be very limited as compared to energy production but it must be stressed that expenditures in this field are difficult to assess. R & D financed under the heading of other energy sectors falling outside the strict definition of "utilisation" (energy conversion for instance) relates in many cases to conservation and serves this purpose.

In absolute figures substantial programmes of R & D undertaken under the heading of conservation and aiming at better utilisation of energy have been planned in France, Germany and the United States. In the United Kingdom, the nationalised fuel and power industries especially have substantial programmes aiming at better utilisation of primary fuels. In percentage of the reported expenditures on energy R & D, France and the United Kingdom devote 7 per cent, Germany 5 per cent and the United States 4 per cent in 1974.

The highest percentages of expenditures are those of Denmark and Finland (14 per cent and 8 per cent) but given the size of their energy budgets, which total less than $5 million, they are not comparable to the lower percentage spent on conservation by the United States. It is questionable however, whether the relation production - conservation is the only criterion for determining the appropriate share of conservation R & D in an energy budget. It is true that for a resourceless country energy economies are in any case desirable, but consideration should also be given to the consumption rate and to the impact of such economies on consumption which is already low.

Although all necessary research in the economic interest of the industries will be undertaken, it remains to be seen whether the initial concern shown by countries regarding conservation will continue to grow or whether the programmes planned at present will slow down as the crisis gradually ceases to attract public attention.

c) The importance of energy carriers

Energy carriers are related to three different aspects of energy R & D. First, there is a direct relationship with energy production insofar as energy carriers ultimately allow its appropriate, efficient and environmentally acceptable conversion, storage and transmission. Second, a type of energy carrier is often related to a particular energy source. Third, for a given energy carrier, there is an inter-relationship between its three functions, i.e. the conversion, storage and transmission of energy.

As regards the relation energy production - energy carriers, the adequacy of research planned on energy carriers compared to energy production is indeed difficult to establish. The effort devoted by different countries is quite uneven and the highest figures - as a percentage of total energy R & D - are those of Sweden (20 per cent), France (11 per cent which includes R & D of nationalised industries), the United Kingdom (9 per cent also including work in the nationalised industries) and Switzerland (approximately 8 per cent). The United States devoted approximately 2 per cent in 1973 and 2 per cent in 1974 which, given the rise in the total R & D budgets from 1973 to 1974, means an increase of 120 per cent against 47 per cent for energy production.

The adequate level of R & D for energy carriers cannot however be set arbitrarily at a given percentage and a meangingful evaluation of R & D expenditures requires a country by country study which takes into account multiple factors such as the development stage of research, the prevailing primary source of energy, geographical conditions, population and industrial distribution.

Furthermore, in observing national programmes and level of funding allocated to different energy carriers, there appears to be no major research on a secondary source system besides electricity, although some programmes on hydrogen and methanol have been announced, and a small effort on district heating is being made by a few countries. For the time being, electricity is more directly linked to the development of nuclear energy but in the long-term other carriers will have to be developed to replace a transportable and flexible fuel like oil and to avoid the considerable heat losses of electricity generation and transmission.

Finally, some imbalance appears in R & D programmes and funding with respect to electricity. Storage seems to receive a relatively low funding as compared to generation and transmission. Although given the stage of research, such funds may prove to be adequate, it must be taken into account that if electricity is to become the major energy carrier, it remains a very difficult form of energy to store for which extensive research is necessary.

d) Energy systems

One of the difficulties confronting governments today in making decisions is the lack of comparative and comprehensive studies providing valid criteria and the basis of a choice as well as the inadequacy of conceptual instruments for assuring that such choice satisfies social and human needs. Major research programmes on energy systems, covering the technical, economic and social components of the energy issue are at present still the prerogative of few countries, namely the United States and at a much lower level of funding, the United Kingdom. These countries devote 1.7 per cent and approximately 1 per cent respectively of their energy R & D budget for 1974. Other countries have announced research programmes on energy systems, namely Denmark, Germany and the Netherlands, but no figures are available.

Such studies are essential for countries having diversified resources and therefore difficult choices to make but they are even more important for countries with scarce resources and average or small R & D potential, as the wrong choices may in such cases be more prejudicial.

o

o o

From the review of the R & D programmes and funding undertaken by different countries and summarised in the preceding sections, some specific trends emerge. Of all the energy sources, nuclear energy constitutes the main R & D objective in most countries. The proportion devoted to it, which in the case of a few countries is slightly lower than 50 per cent, ranges in all the others from 60 per cent to 85 per cent of the total energy R & D budget. Its competitiveness and technological feasibility - at least for fission reaction - and the long-term supply it will provide with the development of fast breeders make nuclear energy the predominant choice despite the controversial environmental problems it raises.

At the other end of the research scale, programmes on new sources, mainly solar and geothermal energy, although very limited in extent and of real concern to a few countries only, are a first sign of a more comprehensive and long-term approach. The percentage devoted to these sources ranges between 1 per cent and 8 per cent except for Iceland where geothermal energy accounts for more than 50 per cent. In absolute figures only Canada and Japan devote an approximate amount of $5 million whilst expenditures in the United States total more than $36 million for 1974.

The only figures available for 2 consecutive years, those of the United States, show almost a fourfold increase but this remains quite an exceptional increase rate limited to one country. Progress will be slow until this field, still mainly supported by governments, attracts industrial interest for research and investment. What seems important, however, is that the research effort on these sources is planned for a number of years (the United States and Japan). The increase planned in the years to come might therefore well bring the research level to the point where their development could no longer be stopped, or at least would no longer be of marginal interest.

Fossil fuels appear in an intermediate position and research expenditures on coal, for instance, are significant only in a few countries; the United States, Germany and at a much lower level,(1) the United Kingdom and France. In these four countries coal R & D for 1974 amounts to approximately $380 million and their percentage is between 15 per cent and 26 per cent of the energy budget of these countries. More difficult to assess is the present research effort on hydrocarbons in the absence of industry R & D figures. Certainly the impact of a dynamic and powerful private industry supporting this field might well speed up the necessary technology for new discoveries and better exploitation of resources.

Although some choices may be considered as a result of decisions taken in a time of crisis and therefore liable to change, the present R & D commitments already indicate the main lines of energy developments for several years to come.

3. The impact of energy R & D on other policies

The extent of the investments for research and a sound energy R & D policy are but one condition for a successful solution of the energy problem within an overall energy policy. Governments will have to make simultaneous efforts on several fronts and undertake research also in connection with energy related policies such as overall policies for science and technology.

Because of the high interdisciplinarity and the complexity of the energy field, progress in energy R & D will in fact depend upon advance of fundamental research in a number of disciplines and in materials science as well as upon the provision of the necessary qualified manpower required to master the resulting technology.

Among the disciplines which should receive increased support, earth sciences - namely geology, geochemistry and geophysics - are of primary importance because of their relevance to the exploration, assessment and exploitation of most energy resources. Research on chemical engineering and particularly on organic and hydrogen

1) Not taking into account industry R & D expenditures.

chemistry and electro-chemistry should also receive adequate attention. A special effort is also required for the development of materials science and technology. Apart from catalysts upon which further progress in chemical engineering depend, research is also needed to improve the economics, the life-time as well as the heat, corrosion and radiation-resistance of materials.

As regards scientific and technical manpower, whilst the sophistication of energy related technologies will require a substantial increase in the supply of engineers, the provision of a great number of technicians will be an equally important condition for the implementation of R & D. A rough estimate for the United States alone indicates that the projected energy programmes require an increase of qualified manpower of as much as 40 per cent by 1980,(1) which means that it will have to be supplied at a rate seven times higher than for the past 12 years.

In a broader social context research on conservation and energy system assumes improvement and progress in social sciences. As it has already been mentioned in this report, the role of social sciences will be to contribute toward the knowledge and understanding of the social obstacles to the public acceptance of certain sources and forms of energy such as nuclear energy or other new energy sources. Social science research is even more essential in the field of energy utilisation. In fact, beyond the energy savings which may be expected from improved production and utilisation technologies, all problems of energy resource conservation are closely linked to the life-style and the industrial and economic structures of society.

Finally, R & D for large scale energy production and utilisation will have to be coupled to research on environmental sciences. Present understanding of the general environmental effects related to large scale production and utilisation of energy is quite insufficient. This is also particularly true for the influence of thermal emissions in the atmosphere. Although the science of climatology is in its infancy and even if it is not known with certainty whether or not a considerable energy increase would have adverse effects on climate at world level, the possible effects on local climates of an important heat flux exceeding solar radiations constitute a serious problem which deserves research and attention.(2)

Moreover, the development and utilisation of all energy resources depend vitally on water and land resources and will give rise to very serious problems of location, installation and land use. The broader issue of the relation of energy to the environment is dealt with in a separate report prepared by the OECD Environment Committee.

1) U.S. Energy Prospects, An Engineering Viewpoint, National Academy of Engineering, Washington D.C., 1974.

2) In the centre of New York for instance, the heat flux is equal to 630 w/m^2 i.e. 9 times the solar heat reflected by the continent.

B. INSTITUTIONAL PROBLEMS

Leading or following the new trend towards a comprehensive energy R & D policy, governments now emphasize the need to create national energy R & D administrations which will be in charge of all scientific and technological aspects of energy. Their task will be to place the numerous disparate energy R & D centres into national, if not international, perspective and to allocate resources to them in accordance with the needs and priorities of the overall policy.

This decision does not always imply the creation of a new agency. In some countries, traditions of decentralisation and the relative inertia of the research system do not lend themselves to a quick reassignment or concentration of R & D responsibilities in new agencies. Some countries, such as Denmark, France, the Netherlands, Sweden and Switzerland, have formed ad hoc co-ordination committees or advisory councils for energy R & D which do not permanently change traditional administrative structures.

In several countries, nuclear energy agencies - to mention the most noticeable example - still have difficulties in finding their place in this new framework. After many years of generous government support and relative independence, nuclear energy agencies do not easily accept being "absorbed" by a broader energy R & D organisation. On the contrary, they occasionally endeavour to become the government organisation responsible for the entire public energy R & D effort themselves.

The creation of a comprehensive energy R & D administration or of a R & D co-ordination committee with similar competences, raises questions about the links between this new body and existing administrations. In principle, the links which can be envisaged for the new body are threefold:

1. with the overall energy policy organisation
2. with the overall science policy organisation
3. with the energy R & D laboratories in industry, university and government.

There is no unique, and probably no ideal solution to the problems of integration between energy R & D administrations, energy policy and science organisations, and energy R & D laboratories. All three links are equally necessary, but depending upon specific needs and traditions, some countries put more emphasis on one type of link than on others.

Energy R & D administrations and energy policy

More or less effective energy policies existed in some, but not in all countries. Already before 1973, ministries in several countries had included specific responsibility for energy policy among their attributes, for example Canada, Ireland, and Turkey. However, in most countries, energy policy was one of the numerous responsibilities of economic, finance or industry ministries.

The energy crisis has so far led to the creation of new ministries and agencies, or other high level institutions, responsible for energy in at least two countries; the Department of Energy in the United Kingdom(1), and the Federal Energy Agency in the United States. Other countries have set up new energy authorities or departments in existing ministries. Some of these countries, e.g. Canada and the United Kingdom, favour the links between energy R & D administrations and general energy policies. Thus, in the United Kingdom, the new Department of Energy has a general co-ordinating role for all R & D linked to energy policy.

Energy R & D administrations and science policy

Other countries seem to lay more emphasis on the links between energy R & D and science policy. They have charged science ministries or corresponding science policy bodies with the administration of energy R & D. This is more than a continuation or reinforcement of the traditional roles of science policy bodies. Science policies of the past usually did not look at "energy" as a whole, but as specific, unrelated, energy sectors. The new task is more global. In Germany, for example, energy R & D programmes are being planned and co-ordinated by the Ministry of Research and Technology, although they are an organic part of the overall energy policy. In Belgium the science policy authorities are responsible for the overall co-ordination of energy R & D.

Energy R & D administrations and energy R & D laboratories

Linking an agency or committee with overall responsibility for energy R & D to the laboratories responsible for executing it is perhaps the most difficult of all three co-ordination tasks.

The autonomy of old, sometimes powerful and mostly resistent research laboratories usually makes close co-ordination with a new administrative body difficult. One way to solve this problem might be to group the administrative responsibility for energy R & D, and the government laboratories carrying out this R & D, into a new, independent institution. Energy research policy and its implementation would come under the same roof.

---

1) The United Kingdom has previously had Ministries dealing specifically with fuel and power; only over recent years had their function been combined with other departments.

The United States envisage trying this solution, apparently the only country to do so for the time being. An Energy Research and Development Agency (ERDA) has been proposed which will be responsible to the President directly, and not to the Federal Energy Agency, nor to the National Science Foundation. However, ERDA will have to seek some form of co-ordination with both science policy and energy policy agencies in spite of its legal independence from other administrations. In the same way, the two other forms of co-ordination - be it administrative integration of energy R & D policy into an energy ministry or into a science ministry - will ultimately succeed only if there are some links with the other ministry as well.

A clear-cut, exclusive assignment of all energy R & D responsibilities to one ministry or administration might facilitate the formulation of national priorities and the carrying out of certain national policies, but it will not guarantee success if the other parts of the administration and of the research system do not co-operate.

The reorganisation of energy R & D centres and responsibilities in the government sector is an indispensable, but not necessarily sufficient condition for the technological solution of future energy problems. Industry is becoming increasingly important in this context, not just as one of the sectors performing energy R & D, but as an active partner in national energy policies. This is largely due to the diversification trend which distinguishes some leading oil companies. Traditionally, they have limited themselves to the production and marketing of oil and oil products, but some of them are now increasingly becoming energy companies in the widest sense. Massive acquisitions of coal and uranium mines and noteworthy R & D efforts in the oil, nuclear, coal, shale, geothermal sectors, have placed a part of the big oil industry in a position of influence and responsibility with regard to all national energy R & D policies. Thus, the industry will have to participate in the formulation and carrying out of national energy R & D policies, if governments want such policies to be fully successful.

II THE STRUCTURE OF R & D IN THE DIFFERENT ENERGY SECTORS

A. THE HYDROCARBONS SECTOR

In this sector, which includes oil, natural gas, shale and tar sands, the activity of big multinational companies has until now dominated most, if not all, aspects including the scientific and technological ones.

During the last few years, oil companies have financed the majority of all R & D relevant to their business, mostly in in-house research. Global figures are not available, but certainly reach several hundred million dollars per year for the seven "majors"(1) alone, while the number of their R & D employees has most probably exceeded 15,000 on a world-wide basis. It is impossible to find out how much of this effort is invested in improving oil prospection, drilling and production technologies, how much goes into other energy fields, such as coal or nuclear research, and how much is chemical research. Industry keeps its R & D projects secret, and very rarely releases R & D results.

In comparison, government-financed R & D programmes have been slim; in no other energy sector is there a similar disproportion between private and public R & D. For the time being, France seems to be the main exception to this rule, at least in relative terms. France spent in 1972 approximately 29 million dollars on petroleum prospection and production R & D, most of it through the French Petroleum Institute (IFP). IFP is a research association which does R & D for industry, keeping in close touch with industry as well as government. Among its achievements is the development of advanced ocean drilling rigs. The United States government spent in 1973 only 18.8 million dollars on oil, gas and shale R & D, geological resource assessment and all other aspects included, but plans to raise this figure to 41.8 million dollars in 1975. Besides the United States and France, Germany is apparently the only country which plans to have government financed programmes of comparable extent. In 1974 Germany is spending approximately 5 million dollars, but will

1) British Petroleum, Exxon, Gulf, Mobil, Shell, Standard Oil of California, Texaco.

increase this amount to about 27 million dollars in 1977. The German government intends to concentrate its R & D programme on deep-drilling technologies which industry has not yet developed.

In spite of these and other possible increases in public R & D expenditures, industry will continue to finance and to carry out the bulk of all oil R & D. Inside industry, the importance attached to R & D and the support which R & D receives from top management, has varied from company to company. This is partly due to the origin of the oil industry. Whereas the pharmaceutical, chemical or electronics industries have grown as "science-based" industries, the basis of the oil industry has not been R & D, but a natural resource. Hence, the oil industry's strength was for a long time based on ownership of oil-wells, of distribution networks and on market and sometimes political factors. Technology played a subsidiary role. Before the Suez crisis of 1956, oil companies did not carry out much petroleum prospection and production R & D. Although the situation has changed since then, some companies apparently still tend to see and plan their future in an economic and political rather than in a scientific and technological context. As a result, their top management does not always give science and technology the attention or the budget which R & D managers deem necessary. Scientific and technological objectives sometimes tend, especially in times of financial stress, to be neglected, because they are basically long-term.

Thus, although the overall oil R & D performance of industry has been very impressive during the past few years, there are probably individual companies where the balance between the economic-political approach and the technological approach to future business could or should be changed in favour of the latter. While this does not in all cases imply increased R & D expenditures, it implies a quicker grasp of new technological opportunities of developing new oil or other energy sources.

The concentration of oil R & D in industry does not mean that all important innovations in the oil sector have come, or will come, only from oil companies. A number of advanced technologies which are, or soon will be indispensable to the oil industry, did not originate in this industry. Certain components of off-shore oil technologies were developed by aero-space companies, originally for other reasons than oil production. Government R & D has played a considerable role in other sectors of technology. For example, the NASA Earth Resource Technology Satellite (ERTS) will make improved prospection for sedimentary basins possible. On the whole, the oil industry depends, perhaps more than in the past, upon the general advancement of science and technology in all relevant fields, and upon the ensuing technology transfers. Of course, this is true for all industries which are undergoing rapid technological change; it is inevitable and underlines the need of close interaction between the oil sector and many other advanced industrial and technological sectors.

Apparently, the university has played a more marginal role than industry in the advancement of the sciences and technologies relevant to oil. While this might be true in purely quantitative terms, university scientists have made quite a number of important proposals and discoveries in the fields of geology, oil prospection, drilling and production and protection of the environment. Some of these ideas and innovations could have a far-reaching effect. Geology is at present perhaps one of the few fields of science where fundamental research and discovery can lead to direct and sometimes immediate industrial application and commercial success, as was the case in organic chemistry and bacteriology in the 19th century. For example, university geologists have recently suggested a new concept of how oil was formed - a problem which is far from being fully understood. Following this suggestion, a company has made successful oil strikes in sites which by traditional concepts were not considered promising.

The oil industry often supports relevant university R & D, and is generally well informed of the work in universities. Industrial generosity has helped to improve and expand university teaching and research in the relevant scientific and engineering disciplines, although it is not certain that this generosity has in each and every circumstance advanced the frontiers of science and technology. Some university scientists in the United States have observed that faculties and universities which received considerable aid from the oil industry, have sometimes felt inhibited to display independent initiative in fields vital to their sponsors.

Universities should probably carry out more independent R & D in fields relevant to the oil sector, especially in geology, geophysics and geochemistry, and consideration should be given to their receiving more government support for this than in the past.

The main conclusion is that the balance between industrial and non-industrial oil research should be to some extent redressed in favour of the latter. This is not a criticism of the oil industry; on the contrary, it is not industry which has funded or carried out too much R & D, it is government and the university which have funded or carried out too little. Government expenditures on oil R & D do not have to match those of industry in order to be effective. A single, well-chosen R & D programme in a critical field can still lead to considerable progress, especially as there is no guarantee that the oil industry can cover all scientific and technological fields which are potentially relevant.

It does not follow from this that governments which finance oil-related R & D must do it in their own laboratories. Many governments prefer to have publicly financed R & D carried out in those industrial or university sectors which have a specific interest in hydrocarbons. However, it does follow that many governments will have to acquire a greater competence and experience in hydrocarbon sciences and technologies. Such competence can be gained in different ways, among which in-house research is only one, though not necessarily the best one for all countries. Without increased competence, many governments will have difficulty in following the often rapid scientific and technological advances in the hydrocarbon sector, and undertaking their policy implications. If they want to support R & D in industry or the university, they will not always be able to make the most useful choices in face of diverging proposals and conflicting claims.(1)

It is true that the industry does not want to lose its traditional pre-eminence in the science and technology of oil. However, its resistance to any government involvement in oil R & D has to be seen within the overall framework of existing relations between industry, government and public opinion - relations which have in several countries become less harmonious than would be desirable.

The success of national energy R & D strategies, and the more rapid improvement of oil technologies, will partly depend upon a better co-ordination and division of labour between industrial companies, and between industry and governments. This implies that the general climate between industry and government should improve, and that the public and private sectors establish greater transparence and mutual communication on R & D.

Several changes are desirable. It is essential that governments make their objectives clear to industry, which has not always been sufficiently done in the past. The oil industry could help by giving up some of its old, more stringent principles of R & D secrecy. This might lead the governments to re-examine the concept of competition applied to the oil industry, at least as far as it relates to R & D.

The secrecy principle results to a large degree from the competitive posture of industry, which governments accepted to maintain because it was supposed to be essential for a cheap and abundant energy supply. However, applied to R & D and even to geological prospection, the classical competition concept and the resulting absence of co-operation between companies and between the private and public sectors have probably become counter-productive. In the medium- and long-term, the supply of energy might be improved if companies were encouraged to collaborate in the development of better techniques, and if

1) A good example of conflicting claims, and of the difficulty of choosing between them can be found, among others, in the various shale-oil extraction proposals recently put forward in the United States by oil companies which all compete for Federal government money.

duplication of certain R & D efforts both within industry and between industry and government were avoided through closer communications and an atmosphere of mutual confidence.

B. THE COAL SECTOR

The R & D organisation in the coal sector differs from that found in the oil sector. Some countries - France and the United Kingdom for example - have nationalised their coal industries, and all or nearly all coal R & D is financed by nationalised industry or subsidised by public funds. Even in countries where the coal sector is still private - the United States, Germany - the majority of all coal R & D is, or soon will be, financed by government funds.

The United States federal government, for example, plans to increase expenditures for coal R & D from 85.1 million dollars in 1973 to 164.4 million dollars in 1974 and 426.7 million dollars in 1975. In the following years, expenditures are expected to remain at the 1975 level. This is probably more than all other OECD countries together - industry, university and government included - will spend on coal R & D. The country which is prepared to make the biggest coal R & D effort after the United States is Germany; it intends to spend approximately 119 million dollars in 1974 and will increase this figure to approximately 164 million dollars by 1977. The Federal government plans to bear two-thirds of these expenditures. France, Japan, and other countries have smaller coal R & D programmes; that of the United Kingdom is under review.

If governments are in all countries the main, if not the only, sponsor of coal R & D, the situation is less uniform at the research laboratory level. In countries which have nationalised the coal industry, all or nearly all coal R & D is carried out by national coal industry laboratories, other public utility laboratories, or government laboratories. Germany has maintained much of its strong tradition in coal R & D, as demonstrated by the work of the research centre of the Bituminous Coal Mining Association (Steinkohlenbergbauverein) in Essen(1) and other R & D centres in industry and the university.

In the United States, the coal industry has traditionally shown little commitment to R & D. Today, the structure of this industry is changing rapidly. The oil industry is buying coal companies in increasing numbers, and it controls already 30 per cent of the known United States coal reserves. Coal mines which are subsidiaries of oil and metal companies, account already for approximately half of

---

1) With an R & D staff of approximately 1,100 employees who cover nearly all scientific and technological aspects of coal, this is still the largest single coal research laboratory in the western world.

the annual coal production of the United States. However, on the basis of all available evidence, it does not seem that these ownership changes have substantially modified the industry's R & D policy. In other words, the oil and metal companies which are buying coal mines, have so far carried out very little coal R & D, certainly as compared to the extent of their overall R & D effort, or to that of their coal production. Hence, coal R & D in the United States is very scattered; Federal government laboratories, a few universities, non-profit research organisations and industrial laboratories are active in coal R & D. In the latter group, we find the electrical industry, the natural gas industry, the engineering industry, electrical power utilities, and increasingly, the aero-space sector, largely because of its materials R & D experience, in addition to independent coal mining companies and oil companies.

The R & D strategy of oil companies in the coal sector appears to be the opposite of what it is in the oil sector. Whereas oil companies wish to bear the cost of their oil R & D themselves and prefer to keep the government out of it, they expect the government to carry almost the entire burden of their own and the nation's coal R & D. This strategy is certainly not unrealistic. It has been estimated that approximately 90 per cent of all energy R & D money which the Federal government intends to spend for Project Independence will directly or indirectly end up in industry anyway.

Both oil and coal research show a disproportion between private and public effort; oil R & D is dominated by industrial, coal R & D by government money. Certainly, industry cannot and needs not to match government expenditures for coal R & D. However, a somewhat bigger independent coal R & D effort by industry would show a more convincing commitment to a field which under any circumstances will need investigation and development for many years before it can modify the energy situation.

On the whole, the increasing domination of coal R & D by government initiatives and expenditures has several positive aspects. Most coal R & D is quite open, partly because so much of it is government financed or subsidised. Many coal R & D laboratories publish, receive visitors and often discuss their ideas and projects, which distinguishes them from the much more secretive oil industry. This openness should facilitate international R & D co-operation, as is the case in no other fossil fuel sector, and few other energy sectors.

The first coal research emergency or crash programmes which appeared during the recent crisis, have not yet fully exploited the opportunities for international co-operation, as they have not been formulated with an eye for these opportunities. It appears today that there are many parallel or similar ideas and projects in coal

R & D. A more systematic exchange of information, the co-ordination and division of labour, perhaps even the redrafting of some national coal R & D plans in view of possibilities of co-operation might reduce costs and lead-times for all participants.

C. NUCLEAR ENERGY

As a means of generating electricity, nuclear energy has been competitive with fossil fuels since 1972, and it is therefore not surprising that, following the quadrupling of oil prices and the 1973 embargo, most OECD countries have decided to increase the proportion of nuclear power in their energy production programmes more quickly than originally intended. Certainly the relatively low cost of nuclear fuel as compared with fossil fuels, the more balanced geo-political distribution of uranium resources and the greater importance of technology in relation to the ownership of the resources themselves are major economic arguments and offer considerable advantages in terms of security of supplies.

The 1973 crisis has thus provided post hoc justification for the substantial amounts spent on nuclear research since the end of the second world war, but it also provides confirmation that the more recent developments were also necessary. Whilst nuclear technology is still changing very rapidly, it is no longer a special research area in which scientific and technological achievements were the main yardsticks of success; nuclear research should be seen in a wider context in which economic, industrial, commercial - but also social - considerations now predominate. The complexity of the nuclear industry and its expected growth, the vast capital expenditure and industrial potential it requires are all basic features governing, to a large extent, the general orientation and structure of national nuclear research programmes.

Countries which do not have adequate resources and the industrial and technological potential for developing their own nuclear industry will have to import most of the technologies involved and their nuclear research effort will consist primarily of keeping pace with general developments in the nuclear sector. Such an effort is in any case essential in order to make it possible to take decisions in the various technological options in full awareness of the facts, to adapt them where appropriate to specific national conditions and to train the qualified staff needed when nuclear power programmes are introduced.

In fact, even for those countries with substantial scientific and industrial potential, the increasing scale of the nuclear sector is more and more often forcing a decision between maintaining the strictly national - or even nationalist - character which has for so long been a feature of technological achievements(1) or opting for diversity in nuclear programmes. For example, because of the very success of the heavy water, natural uranium reactors developed by AECL in Canada, and that of boiling water reactors developed by ASEA-ATOM in Sweden, these two countries are now finding they have to concentrate their efforts on these technologies which are far from being frozen and still call for a considerable amount of research. In most other countries, nuclear R & D is tending to become international, thus opening the door to a wider range of technologies. A substantial proportion of Belgian and Dutch expenditure on nuclear R & D, for example, is for the joint development with Germany of the SNR300 300 MWe fast-breeder reactor. Switzerland is also co-operating with Germany in building a 300 MW high-temperature reactor using direct-cycle gas turbines.

This increasing need for international technological co-operation is also affecting larger countries. Germany's case has already been referred to. In France high-temperature reactors rated at 1000 MW and above are being developed under an agreement between Gulf Atomic International, the French CEA and a consortium of French industrial firms; similarly, the building of the Super Phenix, a 1200 MW fast-breeder reactor, will be a joint enterprise involving French, German and Italian utilities (EDF, RWE and ENEL). A particularly good illustration of the concern for keeping a sufficient number of technological options open by co-operative effort is provided by the co-ordinating group on gas-cooled fast reactor development of the OECD Nuclear Energy Agency, which works in association with an international group of firms concerned with this technology.

The trend towards co-operation is not limited to reactor development; examples are the EURODIF and URENCO organisations for uranium enrichment and the agreement between British, French and German irradiated fuel reprocessing companies to form United Reprocessors GmbH.

Alongside this increasing trend towards the internationalisation of nuclear R & D, particularly in industry, most countries maintain a substantial level of research governed by considerations of national industrial policy. One example is the work being done on fuel element fabrication. There is also research on improvements to light water reactors with the object of replacing licencing agreements by association agreements in which the two partners would be on a more equal footing. Similar considerations are involved in the research on ATR

1) The marked national character of the technological programmes has in fact been accompanied by a considerable increase in international co-operation in the purely scientific aspects of nuclear energy and all questions relating to safety or radiation protection.

reactors in Japan. Maintaining these national R & D programmes does not conflict with the trend towards more intensive international technological co-operation, but is rather complementary in that it aims to introduce a greater measure of give and take into international relations instead of the one-way dependence which has been a feature of the last 10 years.

Up to the present, the United States has been the only country capable of sustaining a broadly diversified nuclear R & D effort on a predominantly national basis. This situation seems unlikely to change during the next few years, but in the longer term serious thought should be given to the possibility of intensifying co-operation with the other OECD countries. A suitable area for such co-operation might well be thermonuclear fusion. Although nuclear fusion is still at the research stage, the cost of any meaningful effort is already high (and increasing), and it seems likely that the substantial resources that will be required at the technological development stage will be a sufficient incentive to stimulate co-ordination if not integration of R & D in this field in the not too distant future.

With nuclear energy now an industrial reality, countries' own efforts in nuclear R & D are being fitted into a broader international framework, and certain changes in the structure of research are becoming necessary. Nuclear research, because of its very nature and the circumstances surrounding its origins, was initially handled by governmental institutions which, largely because of its political and military importance, were given an exceptional role in the countries' research system: AECL in Canada, CEA in France, UKAEA in the United Kingdom, AEC in the United States and so on. However, once progress had been made beyond the stage of scientific and technological innovation and construction of the first industrial plants could be envisaged, the problem arose of setting up a nuclear industry with both the industrial capacity and the scientific and technological potential necessary for it to act as prime contractor in the construction and marketing of nuclear plants.

The success of light water reactors in the United States stems to a very large extent from the fact that from the outset a large share of the necessary research was carried out by industry under contract. Conversely the main problem which French and United Kingdom nuclear policies have encountered was how to develop a well structured nuclear industry, and it is only recently that progress has been made in overcoming this difficulty. Compared with these two countries, Germany had an undoubted advantage in not launching its large-scale nuclear effort until ten years later, by which time its industrial potential was again considerable.

Lastly the customers and users of nuclear power stations, i.e. the utilities, have been obliged gradually to increase their involvement in nuclear R & D in each country. A typical example is the way that the French EDF has become more important in this respect. It may be noted that the growth in international technological co-operation already referred to is partly due to utilities: one illustration is their participation in the SNR300 and Super-Phenix projects mentioned earlier. In the United States, the electricity generating sector is highly fragmented and until very recently played a minor role compared with such large firms as General Electric and Westinghouse. Nevertheless, the establishment of the EPRI (Electric Power Research Institute), in 1973 to serve all utilities and its very rapid growth since that date is a very clear indication of a trend similar to that observed in most European countries.

As regards institutions, nuclear R & D in the different countries has a three-fold basis: governmental organisations (i.e. the Commissions for nuclear energy), the nuclear plant construction industry and the utilities. The success of nuclear research will largely depend upon their co-operation. The balance and distribution of work between these three partners will clearly vary from one country to another depending in particular on the degree of concentration in the nuclear plant construction industry and the utilities sector. Even so, there is no doubt that these two industries will assume increasing importance (though varying from country to country) owing precisely to the increasingly industrial nature of the nuclear sector.

This irreversible trend raises the question of the future of governmental nuclear research organisations. For a number of years, these organisations have already been losing some of the importance they had at the beginning of the nuclear era; in particular they were hit badly by the levelling-off of R & D expenditure at the end of the 1960s. Although this has been somewhat offset by the new impetus that the 1973 crisis has given to nuclear research, it is still necessary to redefine the role of these organisations. The development of non-nuclear activities, however interesting these may be, is only part of the answer, particularly since few attempts at diversification have so far met with conclusive success. The central issue is still the position that these bodies will occupy in the future organisation of the nuclear sector.

There is clearly no question here of giving a final and detailed answer applicable to all OECD countries. Nevertheless it is possible to indicate three broad lines taking account of the special features of the sector and the governmental nature of the research organisations concerned. The first could be to be responsible for long-term research which is too uncertain and too fundamental to be carried out by industry: thermonuclear fusion is at the moment, the most obvious example of this type of activity. It must be added however that industry should be associated with this research at a sufficiently early stage and would gradually take over as progress is made towards development and industrial production.

A second approach could be to maintain research activities aimed at keeping pace with international nuclear developments. This would clearly suit countries without the necessary resources or technological capacity to develop their own nuclear industry, but it is an equally necessary function for all others, too.

A third possibility could be to provide a general view of the whole nuclear sector and to organise co-ordination. The point is that the increasingly industrial character of the nuclear field will lead to greater decentralisation. Among other things this will mean greater efficiency, but at the same time co-ordination and an overall view will become increasingly necessary because of the interdependence and complexity of nuclear systems. Such a function is primarily a matter for a public organisation. In particular it involves identifying gaps and foreseeing bottlenecks as well as initiating the necessary corrective action.

To summarise, governmental nuclear research organisations may find that their function will be redefined; whilst reducing their research activities, especially applied research and development, their role with regard to the framing of policy may be strengthened.

Public opinion is so sensitive to all matters concerning the safety aspects of nuclear installations, radiation protection and radioactive waste management, that governments are faced with another institutional problem, that of the organisations responsible for regulations, and licensing and inspection arrangements for nuclear installations. In most countries, these functions are still shouldered by the atomic energy commissions, i.e. the organisations which are also responsible for developing nuclear energy. For some years now, these organisations have been losing public confidence, not so much for objective reasons but because they are inevitably suspected, being their own judges, of furthering one function at the expense of the other.

It would therefore appear necessary to separate the two functions and to make a different organisation responsible for nuclear safety inspections and regulations, as is already the case in certain countries such as France, Germany, Sweden and the United Kingdom. These organisations must have scientific and technical data, often supplied by the nuclear industry and in particular by the atomic energy commissions, and this could also invite criticism from highly sensitive opinion; for this reason, there seems to be a good case for developing assessment and research centres having maximum independence in relation to the bodies responsible for promoting nuclear energy. The universities should play in some countries a vital part in this respect.

D. OTHER ENERGY SOURCES

Compared to hydrocarbons, coal and nuclear energy, R & D in other energy sources such as the sun - in all its forms - and thermal gradients of the earth has had little support up to now and the organisation of research activities in these sources is just taking shape at present under the pressure of the energy crisis.

This, of course, does not imply that no R & D has been undertaken in the past but on the whole efforts have been fairly scattered and limited mainly to universities and to some governmental departments in the case of solar energy and to a few enterprising utilities in the case of geothermal energy. From 1950 to 1970 for instance, research on terrestrial applications of solar energy in the United States did not exceed $100,000 per year. More federal support has been granted to solar energy utilisation in space, namely for powering artificial satellites, and certainly in a broader sense developments in space research have contributed to advancing knowledge of this energy source. Geothermal energy has been even more neglected, and research in this field is less advanced, although geothermal energy has been utilised for generating electric power or producing hot water for a number of decades.

Present efforts to further the development of these sources of energy and to organise R & D in respect of them must take into account two problems. One is that whilst the technologies related to the utilisation of these energy sources are certainly not new - wind power was harnessed centuries ago - their development as significant energy sources implies nevertheless a new long-term policy approach in government and industry. No doubt industry's early association with the government effort would be desirable but the long-term pay-offs of research and the risks involved will tend to slow down the required effort in the immediate future. For this reason governments will certainly play a leading role in the years to come for the scientific and technological advancement in the field of new sources. The other problem is that most of these sources can be exploited only locally. Because of their remote location - deserts, volcanic regions, agricultural regions - their development will also be dependent on the extent to which an imaginative effort is made for a new concept of an industrial and urban setting responding satisfactorily to these constraints. In a way, the situation today with regard to these sources which lack a set of traditional uses and utilisations like coal and oil, is a rapidly evolving one and for this reason not easy to assess. However, a few promising signs of durable enthusiasm have appeared.

In solar energy, governments are committing themselves more widely to the various direct and indirect applications including heating and cooling of buildings, solar power stations, photovoltaic cells, wind power, ocean thermal conversion and bio-conversion. The main research effort for 1974 is made by the United States with $14 million followed by France ($3.9 million), Japan ($2.7 million), Germany (approximately $1 million in 1974 and $3 millions in 1975), Australia ($0.5 million) and Canada. The United States research effort for solar energy represents 1.4 per cent of the total United States federal energy budget and constitutes a three-fold increase over FY 1973. Furthermore, the 5-year programme "Project Independence" plans to raise the figure to $50 million in FY 1975.

Whereas in Japan the major burden of research is directly supported by government and is carried out in government laboratories, in the United States federally funded laboratories will play an important role. In fact, the research supported by NSF, which represents 85 per cent of the total 1974 effort, will be carried out through subsidies and contracts to a number of universities and national laboratories such as the Jet Propulsion Laboratory or the NASA Lewis Research Centre. NSF support goes also to several industrial laboratories of aerospace companies or electrical enterprises. The remaining 15 per cent of the research effort supported by other United States federal agencies, NASA, AEC, DOD, HUD, is also carried out at federally funded R & D centres such as the AEC's Argonne National Laboratory and the NASA's Goddard Spacecraft Centre. Undoubtedly NASA's role will be decisive for the development of photovoltaic conversion.

As regards the contribution of industry to R & D, there is no doubt that the existing small market for solar energy - as an example the solar cells market accounts for no more than $5 million per year - and the long-term investments needed are hardly attractive to industry at a moment when good opportunities exist for better and quicker returns in the exploitation of other sources (coal gasification, shale processing, etc.). Industry is, however, showing some interest: in the United States some oil companies, utilities and private corporations are funding research (an oil company alone is contributing $3 million) but a more substantial contribution could help develop some particular technologies. Indeed, the ingenuity and forcefulness of industry are essential for the development and commercialisation of existing technologies for some of the applications of solar energy such as the heating and cooling of buildings and solar cells. In such a situation, it is clear that incentives are needed for substantial amounts of additional research.

These may take different forms but the interest of the government itself already serves as an incentive. This seems to be the case of Japan where industry is planning to join the government effort. Other stimulative measures take the form of an appropriate tax policy or legislation. A case in point is a Bill which has been voted in the United States providing a subsidy of $10 million a year for the development of solar heating and cooling and empowering NASA to award contracts for designing and building solar heating and cooling units for houses, factories and colleges. Incentives will have to apply not only to the diversified range of industries involved in the development of solar energy in all its applications but also to architects, executives, consumers, etc.

At the institutional level, R & D on solar energy raises a problem of co-ordination owing to the large number of agencies and departments concerned with the different applications of this source of energy. Ten agencies in the United States and at least five governmental bodies and institutions in France are now engaged in research activities. An attempt to achieve such co-ordination is being made at present in the United States through the designation of the NSF as the lead agency in the field of solar energy and the establishment of an Inter Agency Panel which was set up to this effect in 1973.

Research in geothermal energy shows some analogies with the solar energy field in as much as the majority of R & D is funded by the government. United States, Japan, France and Iceland have allocated for 1974 $13 million, $2.7 million, $1.7 million and approximately $0.8 million respectively. Here the share of the different agencies contributing to research in the United States is more balanced than for solar energy although NSF remains the leading agency for co-ordination of activities. In reality AEC supports 36 per cent of the total, NSF around 30 per cent, USGS, DOD and the Bureau of Mines together 34 per cent. R & D will be carried out, so far as the United States is concerned, at federally funded R & D centres such as the AEC's Los Alamos Laboratory and directly in governmental laboratories in the case of Japan.

In this field too, the research activities of the government are part of a long-term plan. For the United States it is foreseen that an increase of federal funds will bring expenditure for geothermal energy up to $44.7 million in 1975, which is the first of the five years of the "Project Independence" plan. As to Japan, research on geothermal energy is part of the Sunshine project which is to continue up to the year 2000.

This form of energy attracts the interest of fewer industries than does solar energy. Public utilities and some engineering firms have contributed in the past to the development of this source, though public utilities remain the real pioneers in this field. The effort made at present by industry is hard to assess. Large utilities companies in the United States are known to invest more than $15 million for developing geothermal units and are undertaking R & D in connection with the tapping of hot brines. In addition, oil companies have lately shown some interest. Nevertheless it is recognised by governments that industry R & D should be encouraged to contribute more vigorously to knowledge in this field. The government effort certainly encourages R & D and for instance in Japan, at least 5 of the 10 regional utilities companies have put or are putting a major emphasis on electricity derived from geothermal energy, while research committees have been created in most utilities companies in response to the Sunshine project. However the risks inherent in the drilling, location and lifetime of geothermal reservoirs still account for the slow increase of industrial R & D. Here, too, therefore, various incentives are necessary if a demand large enough to stimulate research is to be developed.

In conclusion then both for solar and geothermal energy, the development of technology will certainly require industrial involvement but developing technology will not be enough. A parallel effort has to be made to encourage utilisation of these technologies through various measures such as tax incentives and the multiplication of demonstration plants and projects. Further, because these sources represent alternative ways of supplying energy which are desirable for a number of countries and as they remain at present at the lowest level of funding, an internationally co-ordinated research effort may bring about desired results without imposing too heavy a burden on any single country.

E. ENERGY CARRIERS AND RELATED TECHNOLOGIES

One obvious conclusion from a study of Member countries' research programmes on energy carriers is the existence of a very marked imbalance between the substantial amount of work being done on these energy carriers that are now in use - oil products, natural gas and electricity - and the extremely low level of activity regarding other possible carriers. As a result of the 1973 crisis, arising from a relative and temporary scarcity of oil, one of the primary energy sources, the diversification of such sources became an objective of paramount importance, but it did not provide a similar stimulus to the search for other energy carriers despite the fact that energy carriers are the keystone of energy systems.

Quite apart from the intrinsic advantages and drawbacks of different energy carriers, this imbalance is due largely to the existing structure of the field of energy carriers which does not lend itself to the development of research that would enable new ones to be developed.

As far as oil products are concerned, the outstanding feature is the extensive vertical integration, in the oil companies, of the primary and secondary agents. This is a well established sector but considerable technological progress has been achieved during the last few years. Over the last ten years, research by the oil companies has been concentrated more on energy carriers and on petrochemicals than on the exploration of new fields. In particular, significant progress has been made in the technologies of converting crude and in solving the related environmental problems.

The integration which is a feature of the oil sector does not apply in the case of natural gas. The distribution companies buy most of their gas from the oil companies, since natural gas and oil come from similar geological formations and are often found together in the same deposit. Another difference is that the gas firms are generally confined within national boundaries, and in many cases nationalised; they are always considered as a public service and are thus subject to governmental control. Most of the research carried out by these firms concerns distribution system technology and management.

Research in the electricity sector, as already indicated in relation to nuclear energy, is carried out primarily by the electrical equipment industry and the utilities which, like the gas undertakings, are public utilities. The respective parts played by these two industries vary from one country to another, but they are invariably closely linked. Over the last fifteen years or so, the relationship has been carried over into the nuclear sector and the electrical equipment industry and the utilities are thus tending towards greater vertical integration, ranging from the conversion of the primary energy source to the production of electrical equipment.(1)

The electricity sector, being newer and more science-based than the two other existing energy industries, is the most dynamic of the three from the technological standpoint. By and large, the three main branches - generation, transmission and storage - are covered in a fairly balanced way. It should also be noted that the utilities have carried out substantial research into the structure and management of electricity systems and that the experience thus gained could prove extremely valuable for research on overall energy systems.

1) At the same time, there is a trend towards greater horizontal integration of energy sources as the large oil companies develop their coal, uranium and even high-temperature nuclear reactor activities.

The three existing types of energy carrier are in competition at the energy utilisation stage although only a limited number of applications are involved and the changeover from one energy carrier to another is necessarily gradual since it is dependent on new equipment. In other respects, the three sectors have remained highly independent of each other. Their undoubted technological vigour has rarely gone beyond the boundaries of the type of energy carrier for which they were responsible. Of course there are noteworthy exceptions, one example being the considerable amount of research carried out in Sweden and Germany into the use of power station waste heat. In a few countries, e.g. France and the United States, the gas industry has extended its research to cover gas produced from sources other than natural gas and to the use of gases other than methane, notably hydrogen. Although important, these examples are only exceptions and in no way affect the beginning of this section.

This paucity of research into possible new energy carriers mainly reflects a shortcoming in energy policies comparable to that which, prior to 1973, led to insufficient R & D being devoted to energy sources other than oil. For this reason, energy carriers should be given special attention by governments and particularly by the organisations responsible for energy R & D.

It would seem particularly necessary to include in this context research dealing with existing energy carriers, to extend research on other possible energy carriers and to make a comparative study of the respective advantages and disadvantages of the various energy carriers in order to decide how best they may be combined. As far as new carriers are concerned, one of the essential tasks of the energy R & D policy institutions or mechanisms would be that of coordinating research into the production and utilisation of such carriers.

The question of new energy carriers goes far beyond the scientific and technological aspects. Quite apart from the R & D they merit, there is the more delicate problem of incorporating them in the energy systems, and that of overcoming the natural resistance of established practice and interests.

F. UTILISATION AND CONSERVATION OF ENERGY

Until 1973, with a few exceptions, a feature of energy R & D in most countries was the dearth of any research aimed at saving energy through rationalising its use. The 1973 oil crisis drew attention - rather belatedly - to the extent of this imbalance and, to judge by all countries' declared intentions, energy saving now seems likely to become one of the main objectives of future energy R & D policies. Whilst a substantial increase in R & D spending for this purpose is certainly necessary, it is not the complete answer. Policies in this field will not produce significant results unless they take into account all the technological and institutional factors involved in the use of energy, which call for very different procedures from those applicable to energy production.

When energy reaches the stage of utilisation, it is only one factor - though a fundamental one - in technologies covering a variety of functions (transport, housing, production of goods and so on); what is more, other factors, such as the environment, material resources and questions of labour and capital, have to be considered as well. Another basic fact to be taken into account is the number and extreme diversity of the technologies involved in the use of energy. Although it is possible to pick out certain energy-intensive technologies (e.g. motor cars, aluminium smelting and steel production), it is essential not to underestimate the part played by the sum total of a large number of micro-technologies, which are much more difficult to identify and evaluate, and whose impact on energy consumption is little known.

The diversity of the sectors using energy is just as wide as that of the technologies involved. A division into broad sectors such as the domestic and commercial sector, industry, agriculture, and transport is no more than a first approximation and considerable differences exist within each of these sectors. In the domestic and commercial sector, for example, the construction industry is totally unlike the household appliance industry, and the behaviour patterns of an individual consumer and a supermarket are obviously governed by different motives and constraints. The industrial sector includes not only large but also small- and medium-sized firms for whom the question of energy utilisation obviously arises in very different forms, and so on.

Faced with this variety of technologies and structures, government scientific and technological policies relating to the use of energy will therefore have to be highly flexible in their application. Each case will have to be treated on its merits according to the technologies and structures of each specific sector of utilisation. Flexibility is particularly essential since it will have to go hand in hand with far-reaching decentralisation in the planning and performance of research.

It is a striking fact that there are many technological ways of saving energy which, although available, have not yet been applied on any large-scale. The reasons for this situation are only partly economic (e.g. the low cost of energy prior to 1973, and problems of financing investment). As regards research and development, the moral to be drawn is that scientific and technological policies in the field of energy utilisation should not merely increase the quantity of research, but should first endeavour to steer research towards the widest concrete application of its results. In other words, these policies will undoubtedly fail unless they are primarily innovation policies.

From experience in other scientific and technological fields, it is clear that governmental action will the more effective as the need for research and innovation is more clearly realised by the people and organisations making up the various utilisation sectors, and as these people and organisations become more closely involved in planning and conducting the research. As has already been said, it is self-evident that the methods used will have to be chosen according to the specific circumstances in each sector. Regulations making for greater energy efficiency may, for example, occasionally prove as effective as financing research; in certain cases the research could be carried out by the firms that use the energy, while in others the industry responsible for energy carriers may prove to be more suitable. Sometimes government laboratories may be more appropriate. In any event, whatever method is chosen, participation by the user sectors is essential if research is to be effective.

Regardless of the flexibility and decentralisation that are necessary for applying R & D policies relative to the use of energy, it is not possible to regard these policies simply as the sum of the programmes concerning the various user sectors. To begin with, the utilisation of energy must obviously be viewed in relation to the other aspects of the energy field - energy carriers in particular. Secondly, as already stated,(1) the limitations on progress made possible by improvements in individual technologies cannot be overcome unless research deals also with the various energy utilisation systems: urban, industrial and transport systems, etc. The very nature of this kind of research, its relationship with other aspects of government policy, and the fact that it must be co-ordinated with work on individual technologies, means that the framing of R & D policies on the use of energy calls for the sort of coherence and overall viewpoint that are only possible with R & D policies covering the whole field.

1) See Part II, Chapter V.

III OPPORTUNITIES FOR INTERNATIONAL CO-OPERATION

The serious limitations, and in some cases failures, of international co-operation in R & D which have occurred in the past in some specific sectors, force recognition of the fact that international co-operation will succeed on condition that it is inspired by a common political will and policy of the partners involved and not only by the internal logics of science and technology or that of the institutions created for the purpose. Besides, not all forms of co-operation have proved to be equally efficient, and in this regard bilateral and trilateral agreements have met with less difficulties than broader international co-operation schemes.

However, before examining the problem of international scientific and technological relations in the particular case of energy, it may be useful to review briefly the advantages that could be gained in general by international co-operation in R & D:

- it makes greater resources available, in terms of the information, knowledge and know-how necessary for any R & D activity;
- it means that the whole process, from the research stage right up to the practical application of the results, may be speeded up;
- it can reduce the overall cost of R & D programmes through more efficient utilisation of financial resources, equipment and skilled personnel;
- it makes possible a wider range of approaches.

There are also the following specific factors which render scientific and technological co-operation not merely desirable but essential in the field of energy:

- the importance of what is at stake: the economic and social structure of OECD countries is dependent upon the intensive use of energy; this situation cannot suddenly be altered without grave consequences for national and international equilibrium.
- the international dimension of all aspects of energy.(1)

---

1) cf. Part I

- the vital importance of research: during the sixties, oil dominated the scene because of the exceptional combination of advantages it offered; being cheap, plentiful, and easy to transport, store and use. Now that some of these advantages have disappeared, the need to find other, substitute sources, makes it essential to undertake systematic, large-scale research and development covering the totality of the energy field.
- the extent of the energy sector is such that no single country can carry out an R & D programme on this scale.
- the increasing complexity of the field: in the past the only determinant factors were technical and economic, but now the importance of social and environmental aspects must be taken seriously into account.

As in other areas, the development of international co-operation in energy R & D is largely dependent on development of the national R & D effort, of which it is merely an extension. Conversely, in the formulation of national R & D policies covering the whole field, possible ways and means of co-operating with other countries, and the requirements of such co-operation, should be systematically surveyed. Thus, before identifying principal areas for research that should be given priority in any joint action by OECD countries, a review of existing co-operative activities in the various sectors of the energy field would seem to be necessary in order to identify the areas where increased co-operation appears desirable and feasible. Consideration of activities in co-operation should clearly not be confined to the scientific and technological relations between the OECD countries. It should extend to co-operation between developed and developing countries; in particular, the latter should be associated to co-operative programmes at their very beginning. However, since this question arises in a different context, it is dealt with in a different chapter of this report.

A. THE PRESENT SITUATION IN THE DIFFERENT ENERGY SECTORS

International relations in the energy field are already well-established, as evidenced by the number of bilateral agreements and international organisations to which reference has already been made, to say nothing of the informal contacts between scientists and laboratories and the agreements between industrial firms. Nevertheless, these international links may be misleading: in many cases all that is involved is the exchange of general information, rather than a true co-ordination of research projects, or research carried out in common. In particular, the existing pattern of international relations is very uneven in coverage and highly dispersed. It may be noted here that the weaknesses of international co-operation very largely reflect those of national policies.

Little information is available on research carried out by industry and this report therefore deals mainly with sectors in which governments play the predominant role; generally speaking, these cover medium and long-term research or research bearing on environmental and safety aspects. In the case of industrial research, the initiative regarding co-operation is usually left with the firms themselves and the role of government is primarily to provide information about the possible areas for such co-operation and to remove any legal or procedural obstacles. However, this does not mean that governments could not assume a more positive role in promoting co-operation in relatively short-term research; one possibility which would appear to deserve careful consideration would be for several countries to join together in the financing of demonstration plants.

1. Fossil fuels

Co-operation in this sector especially as regards oil and natural gas, is relatively limited, or at least little is known about it, the industries concerned being mainly in private hands. As far as exploration and resource evaluation are concerned, there is a measure of world-level co-operation in theoretical questions within the International Union of Geological Science. At the more practical level, the heads of the geological research organisations in the different European countries meet once a year to discuss problems of common interest. Co-operation between Member governments on this question should be far more intensive and systematic in view of its importance in the formulation of energy policies.

The oil and natural gas production sector could be a suitable area of co-operation, but the fact that it is mainly in the hands of private industry may limit opportunities for governmental co-operative actions. On the other hand it may to some extent be regarded as international in view of the predominant role of the multinational oil companies. However, not much is known about the relations between these companies and their national subsidiaries, or about co-operation between the various multinational companies. One example of this type of co-operation is SEAL (Sub-sea Equipment Association Limited), an engineering firm set up by CFP, BP, Mobil, Westinghouse and DEEP (a group which includes the Institut français du pétrole) and specialising in offshore oil.

Although private industry largely dominates this area, governments could still fulfil an important function in developing co-operation in certain technologies, such as offshore oil research, the underground or sea-bed storage of hydrocarbons and the transport of gas by sea. In a more general way, they could encourage the gas distribution companies, which are public utilities (and, in many countries, nationalised undertakings) to do more about co-operation. Organisations already existing include the European Gas Research Group, of which the gas undertakings of six of the Common Market countries are members, and the Gas Atlantic Research Exchange formed by Gaz de France, the British Gas Corporation and the American Gas Council.

More information is available about international relations in the coal sector, where governments have greater responsibility. The West European Coal Producers' Association (CEPCEO) and more especially the European Coal and Steel Community (ECSC) continue to provide an institutional framework for co-operation of a most extensive nature. Co-operation regarding safety in coal mines also has a long history: international conferences on this subject have been held every two years since 1931, the most recent meeting having taken place in 1973. In addition, the heads of R & D departments concerned with safety questions in Belgium, France, Germany and the United Kingdom meet annually to discuss and co-ordinate their plans. Representatives of the United States Bureau of Mines have attended on the last two occasions.

A major event as regards R & D aimed at improving coal combustion and conversion processes has been the conclusion of bilateral co-operative agreements between Germany and the United States, and between the United Kingdom and the United States. The ECSC has put forward plans for future co-operation in these areas.

Thus there is already a substantial measure of co-operation in the coal sector, but even so, in view of the renewed importance which coal seems likely to assume, and the research programmes now under way in many countries, co-operation needs to be systematically extended and intensified, with particular reference to mining, utilisation and conversion techniques. Here governments must be the prime movers since they play a key role in the sector at national level, but industry must be fully involved in any plans for co-operation.

For resources such as oil shale, tar sands and peat whose importance is confined to a small number of countries, the countries concerned might well find it worthwhile to consider co-operation in the form of bilateral or trilateral agreements.

2. Nuclear Energy

From the outset, nuclear energy has been the subject of particularly intensive international relations. Although, there are in this sector, more than in any other, very many bilateral agreements and international organisations (IAEA, NEA and EURATOM), there is no disguising the intense rivalry that exists between countries, and the fact that co-operation could be appreciably improved in many respects.

Reference was made above(1) to the trend toward increased internationalisation in industry. This is a logical feature of development in the nuclear sector and, far from obstructing it for reasons of short-term national interest, Member governments should give it every encouragement.

Other important aspects of nuclear energy calling for greater co-operation between governments are all those questions relating to the environment, safety and radioactive wastes. Radiation protection, for example, is probably the field where the record of co-operation has been most outstanding, as a result of the work of the International Commission for Radiological Protection whose function is to lay down radiation protection norms and principles, the problem of application being dealt with by the IAEA, the ENA and the European Communities. These same organisations are actively concerned with the safety of nuclear installations, an area in which there are also many bilateral agreements. Arrangements for exchanging information are already well developed, but more needs to be done in terms of the international use of national research facilities and the construction of international installations. Systematic international co-ordination is also required in radioactive waste research. The new EUROCHEMIC programme on waste management, which is to start in 1975, holds considerable promise.

Medium and long-term nuclear research is another area where governments would be well advised to intensify international co-operation. Three examples of the subjects involved are the direct utilisation of nuclear heat, nuclear power units for ship propulsion (with particular reference to questions of safety and port installations) and, above all, controlled thermonuclear fusion. Co-operation in this field, which will continue to be more scientific than technological for a considerable time to come, is already well-established; an international scientific community already exists, somewhat similar to that in disciplines such as astronomy or high energy physics. Moreover, virtually all fusion research in the EEC countries is incorporated in a co-ordinated joint programme. In the future, this co-operation could well be extended to all OECD countries, especially for the construction of future generations of large facilities of the Tokamak type. Co-operation could already be stepped up with regard to certain technological problems such

1) cf. Part III, Chapter II

as the effects of radiation on the materials used in fusion reactors, and facilities such as high intensity neutron sources could be built on a joint basis.

3. Other energy sources

Because of the small amount of research devoted to other energy sources, international co-operation in this area is still extremely limited, often consisting purely in personal contacts between specialists in the various research topics involved. For some years, there has been a solar energy programme at the European Communities joint research centre at Ispra. Recently NATO has undertaken to co-ordinate research in its member states on solar energy, deep-sea thermal gradients and geothermal energy.

This co-operation needs to be considerably intensified and extended, but in general it is the overall question of alternative energy sources which requires an international approach. Apart from increasing the number of research projects undertaken jointly by groups of countries, such co-operation should include the systematic evaluation of the energy potential of these sources and provide the necessary machinery to enable the many projects proposed by specialists in the various countries to be examined at international level.

4. Energy Carriers and related technologies

Electricity research is mainly carried out by the manufacturers of electrical equipment and the electricity generating and distribution undertakings. The initiative for expanding international co-operation in this sector must therefore lie with the firms and public utilities concerned. Governments could nevertheless promote co-operation in medium and long-term research, where their role is often more important. One example is MHD for which a working group, set up jointly by ENA and IAEA, already exists. Another example is the case of other advanced cycle systems. The technology of superconductors for the storage and transmission of electricity is another area where co-operation could be intensified. For the storage of electricity, other important topics suitable for co-operative research are batteries.

However, the sector in which governments should take the initiative is co-operation in other research on energy carriers, although it is essential to involve the firms concerned as far as this is possible. Areas in which co-operation is at present minimal and where special efforts should be made in view of the new interest being shown in most countries include the utilisation of waste heat from power stations, combined plants producing both heat and electricity, and heat transport and storage.

Many countries have now begun research on hydrogen and, more generally on the carriers which would enable heat to be transported in the form of the energy of the chemical bond; co-operation is also advisable here. There is now a programme on hydrogen at the European Communities joint centre at Ispra, but much wider and more systematic co-operation is desirable.

5. Utilisation and conservation of energy

Simply because of the total lack of research on the subject in most countries it is this area where international co-operation is at its lowest level. However, NATO recently organised a seminar on the question and is at present planning to give it greater importance. A very broad basis of co-operation would seem necessary since this research sector is likely to expand in all countries.

It must be borne in mind, however, firstly that most of the research concerned is carried out by industry, and secondly that it is often concerned with "microtechnologies". The main role of governments should therefore be to establish a systematic and very broadly based exchange of information and co-ordination machinery covering all sectors of energy utilisation. Once such a mechanism is set up, it should be possible to identify those research programmes that could best be carried out jointly by a number of countries. Among these, co-operation on development of heat pumps and heat exchanges could be undertaken. An important point to remember here is that the research should cover not only utilisation technologies but also systems of utilisation. It is the latter which perhaps lends itself more easily to co-operation between governments.

6. Energy systems

Joint international research is already being carried out on three aspects of energy systems. The application of systems analysis methods and techniques to energy systems is being studied by the IIASA (International Institute for Applied Systems Analysis); energy economics, and in particular the problems of forecasting supply and demand are well-established subjects of co-operation within the OECD; lastly the many effects that producing and consuming energy has on the environment - in the widest sense of the term and including, for example, its effects on health - are being studied directly or indirectly by a number of international organisations (e.g. the United Nations programme on the environment, the World Meteorological Organisation, the World Health Organisation, the IAEA, the OECD and the European Communities).

Research on energy systems is important in the formulation of energy policies and co-operation between governments in this connection needs to be stepped up and to become more systematic. Research itself should also pay due regard to the fact that energy systems are intrinsically international.

B. SOME PRIORITIES FOR INTERNATIONAL CO-OPERATION

International co-operation in R & D must be viewed as an extension of national R & D and national efforts should be an integral part of overall energy R & D policy. Consequently, any joint action with regard to possible co-operation topics, cannot be planned in isolation and must be fitted into the broader framework of international co-operation with respect to energy R & D policy.

There is no question, however, of attempting to deal with all the topics or sectors mentioned above at once. It has therefore appeared necessary to select certain research fields where it is clear that there is an urgent need for greater co-operation between the OECD countries and where this can be achieved relatively quickly. The selection has been based on four main criteria:

- The quantitative or qualitative relevance for energy production, energy conservation or the formulation of energy policies, of research on the subjects envisaged. The extent of this relevance has been evaluated on the basis of the scientific and technological possibilities considered in Part II. It must also be borne in mind that, because of the very nature of R & D, it is by no means a foregone conclusion that its results will be successful; any contribution it may make can only become effective in the medium and long-term.
- The unquestionable advantage of a more broadly based co-operation, as shown up by the review of the present situation in the foregoing paragraphs. In particular, subjects where co-operation already appears to be satisfactory have not been included, although of course improvements are always possible.
- The absence of major obstacles, whether political, institutional or economic, that would prevent co-operation from being improved. Co-operation is easier in research than in development activities because of the industrial and commercial interests that are involved in the latter. For the same reason, preference has been given to subjects in which governments have a major part to play, although this does not necessarily mean that industry is excluded. In applying

these criteria, consideration has been given to the kind of research involved, and to research policy and organisation in the different countries, as analysed in Chapters I and II above.

- The existence of research programmes on a big enough scale in a number of Member countries. It is clear that increased co-operation is dependent on the extent of national activities. The scale of these programmes also provides some indication of the degree of interest of the countries concerned. The programmes in the various countries are listed in Annex III and summarised in the tables, and the criterion was based on this information.

According to the above criteria, at least fourteen fields, as indicated below, can be considered as most suitable for immediate international co-operation. Obviously, for each subject, the mode and type of co-operation will have to be determined, taking into account the framework in which such co-operation will develop, the R & D performance and potential participating countries and the most adequate procedures and mechanisms. The suggested subjects are the following:

1. Exploration and evaluation of energy resources
2. Underground or sea-bed storage of fossil fuels
3. The coal sector (mining, combustion, gasification and liquefaction techniques)
4. Safety of nuclear installations
5. Radioactive waste management
6. Thermonuclear fusion
7. Solar energy
8. Geothermal energy
9. Energy produced from industrial and urban wastes
10. Fuel cells
11. The use of heat as an energy carrier (covering not only waste heat from power stations but also the transportation and storage of heat)
12. New energy carriers (methanol, hydrogen and, more generally, carriers enabling energy to be transported in the form of the energy of the chemical bond)
13. Energy conservation (rational use of energy) in the residential, industrial and agricultural sectors.
14. General studies on energy systems.

Apart from a few differences in the number and definitions of the above subjects, this list includes all and more topics than the ones contained in the report of the ad hoc Group on International Co-operation on Energy Research and Development of the Energy Co-ordinating Group (ECG/ERD/36 final dated 6th June 1974) set up as a result of the Washington Conference in February 1974.

This similarity is easily explained: the nature of scientific and technological problems does not depend on where they are examined; the subjects have been chosen by a similar process; the criteria used are equivalent if not identical; the data on national R & D activities are the same; finally, there is a common basis of expertise.

It is clear that drawing up a list is only an initial stage and that it needs to be followed by detailed discussion of each subject by specialists from the countries concerned in order to determine which subjects under the different headings might be tackled on a joint basis and what form this co-operation should take. The report of the ad hoc Group of the Energy Co-ordinating Group referred to above would form a very useful starting point for these discussions, particularly where the subjects chosen by that Group are the same as those selected by the OECD Committee for Scientific and Technological Policy.

With no intention of anticipating the outcome of these discussions, the following remarks are offered concerning the forms which co-operation might take and which must be tailored to the scientific, technical, institutional and political aspects and constraints of each subject. Whilst in many areas it would be beneficial to move ahead with co-operative R & D projects, in other cases it may be better to co-ordinate national efforts than to institute joint research programmes or set up joint facilities. Finally, in some instances, a most appropriate form of co-operation, at least at the outset, may turn out to be a systematic exchange of information. The degree to which industry will be involved may also vary from case to case.

Clearly the mecanisms adopted should not only make maximum use of existing international organisations but should also permit special "à la carte" co-operative schemes. It is clear that not all Member countries will be interested in every subject and a relevant point here, is that the OECD provides a particularly suitable framework for co-operation of this kind.(1)

The reseach topics in the above list are still only one aspect of the co-operation that is called for; joint arrangements are also needed for the training of scientists and engineers. As already pointed out,(2) this is one of the keys to success in energy R & D. Scholarships enabling young scientists to train abroad, exchanges of scientists, and refresher courses and advanced-level seminars open to scientists from other countries, all need to be developed systematically.

---

1) A provision exists in the budget of the Organisation (Part II of the Budget) which allows for co-operative schemes between only those countries which are interested in a given subject.

2) Cf. Part III, Chapter I.

It is worth stressing that whilst OECD countries may not be able to cover individually the whole field of energy research on their own, their co-operation in specific and different topics of their choice will enable them to be active jointly on all research fronts. This is not merely a desirable objective; it is essential and means that work will have to be distributed in the best way possible.

The purpose and scope of this report was not to make a close and very detailed examination of the possible co-operation areas, and the above suggestions represent therefore an incomplete listing in general terms of the possible themes of co-operation between the OECD countries. Further examination of suitable fields for bilateral and trilateral co-operation should be undertaken by individual countries using, for example, as a basis, the list of national programmes given in Annex I.

It is also important to underline that the approach adopted here, based on research needs and on present and future programmes, need not be exclusive. A group of countries that has decided to pursue a joint energy policy may find it more appropriate to approach the question of scientific and technological co-operation by identifying the research sectors and type of organisation on which this common energy policy might best be based.

IV ENERGY R & D AND THE DEVELOPING COUNTRIES

The following section is intended to present a brief general review of the prospects and problems of energy R & D in developing countries. Contrary to the chapters related to OECD countries, the analysis is not in this case based on countries' replies and for this reason it does not offer a sufficiently full and detailed appreciation of the energy situation and problems in those countries. It should therefore be considered only as an attempt to give a general view of the problems involved and preliminary suggestions for co-operative actions.

Although they account for nearly 70 per cent of the world population, developing countries consume less than 15 per cent of world oil production; in other words for them, unlike the developed countries, the "energy crisis" could hardly be said to create a really new situation; it merely aggravated their existing difficulties(1) which in some cases have now become critical. The change in relative prices following the oil crisis did no more than worsen trends that were already there. An UNCTAD report dated 4th April, 1974(2) shows that during the world inflation between 1972 and 1974 the developing countries have been harder hit by the increase in prices of food products, fertilizers and manufactured goods imported from developed countries than by the higher cost of oil. Fertilizer and food product prices, in particular, began to rise before the prices of oil were jacked up.

Nevertheless the escalation in the cost of imported energy, coming on top of that in the cost of industrial and food products, is having very serious consequences for the already fragile economies of those countries in the Third World which do not produce oil or gas. The balance of payments is disturbed, growth rates held back,

---

1) This does not, of course, apply to the oil-producing countries which have, on the contrary, profited from the "oil crisis".

2) UN Conference on Trade and Development: Problems of Raw Materials and Development: Note by the Secretary-General of UNCTAD. (UNCTAD/OSG/52 and UNCTAD/OSG/52/Add.1).

unemployment made more acute, levels of consumption including that of food, reduced and the degree of dependence intensified. Today economic development is in a worse situation than ever.

Energy problems in the Third World must be considered jointly with the situation in developed countries whose extensive requirements will, among other things, affect availabilities in resources and equipment. Another point to be borne in mind is that, with the present crisis, there is a world tendency to relocate some heavy industries now in the developing countries close to the points where energy and raw materials are produced.

Above all, the developing countries are in a position of dependence. Apart from the fact that those countries not producing oil have to rely on others for their supplies and finance there is the general technological dependence on the developed countries. Recourse to the latters' scientific and technological potential is unavoidable and essential if the developing countries are to overcome the problems that the present energy situation is creating for them; but in deciding when technologies should be developed and how they may be adapted, full account will need to be taken of the specific energy requirements and conditions in each country.

A last point is that the developing countries are far from forming a homogeneous bloc. This is not only true for types of economic growth and for the "level" that they have reached; it is also true for energy. Energy resources and requirements vary considerably from one country or group of countries to another. As far as the possible R & D options are concerned therefore, all that can be done here is to give a few general indications which would then need to be re-examined and defined, as required by the particular situation in each country.

It should be noted, at the outset, that little information is currently available on the energy requirements of the developing countries and that a systematic study of the subject would be necessary before any decisions could be taken with regard to the technologies that might help to meet these requirements. This report is not the place for any detailed consideration of the question but a few general indications are given below which could serve as a basis for the studies that would need to be undertaken.

In the developed countries it is usual to classify energy consumption by three approximately equivalent user sectors:

- domestic and commercial
- industry
- transport.

This breakdown would appear to be inadequate in the case of the developing countries for which a combination of two complementary types of classification might be suggested, as follows:

- the first would be a breakdown by sector, covering industry, agriculture, transport and the domestic sector;

- the second would have a geographical basis, breaking requirements down by urban and rural areas. The domestic and transport sectors would be found in both types of area but industry is generally specific to urban and agriculture to rural areas.

As regards industry, the extent and nature of energy requirements vary with industrialisation policy and depend on whether the primary objective is to cover the basic requirements of rural populations in order to help them reach an acceptable level of living conditions, or whether the aim is to gain a foothold in the world market. Transport is an extremely important sector because of the very large areas involved in many developing countries. A major difference from most developed countries is that the private car has a minor role compared with public passenger transport and goods transport.

Requirements in rural areas are larger than is generally thought, sometimes totalling even more than those for industry, and are highly varied in nature. The population is generally widely scattered in these areas and transport and communications facilities are lacking. The requirements relate mainly to agriculture. At the present time any growth in agricultural production necessarily implies an increase in the consumption of energy in two main forms: firstly the use of fertilizers and pesticides and secondly the use of agricultural equipment (in the broad sense). To this must be added certain specific needs, for example, pumping water for irrigation.

Agricultural development is a crucial problem in all developing countries. It is as important as the industrialisation problem. Through agricultural development it should be possible to feed the population (and thus avoid having to buy food at astronomic prices on the world market), provide work for the large number of workless thus alleviating rural poverty, and possibly produce a surplus that might be sold to build up capital in other sectors. The problem lies in the fact that agriculture is a traditional and backward sector relying on outdated and unproductive techniques. This is, however, a challenge to be taken up by any development policy, particularly that relating to energy.

In addition to the direct requirements of agriculture there are those of any small artisanal enterprises that may exist in the country areas and those of the peasants themselves, however marginal these may be in both cases. Lastly there are the requirements of the community: communications, public health and education, telecommunications, etc.

Further detailed study would be required under each of these general headings in order to establish the real extent of energy requirements and their breakdown in space and time in relation to the differences between rural communities, nomadic communities, towns and industrial areas, etc.

From the scientific standpoint, the technologies and resources needed by the developing countries to meet the energy requirements referred to in the preceding paragraphs are no different from those that are now the subject of research in the OECD countries. The main problem is therefore to select those technologies and energy sources that best suit the specific conditions in the developing countries themselves. These decisions will obviously depend upon physical factors such as climate, topography, natural resources, and so on, and on the overall economic development strategy that is being followed.

In some cases, for example, investment could be centred on a "heavy" energy sector and in others distributed as evenly as possible between regions and sectors. These two options have radically differing implications in terms of way of life of the population, degree of dependence on other countries and exploitation of local resources. In practice the two types of policy are sometimes combined.

The yardsticks for decisions on choice of energy sources and technologies include: variety of potential usages (covering the requirements of both urban and rural areas), cost minimisation, readiness dates, safety and environmental considerations. The main requirement however, is that the choice of technology should be made in relation to developing countries' own raw material and manpower resources. Many of the energy production and conversion technologies in the developed countries call for very large capital investment and are based on increasingly mechanised and automated equipment calling for highly qualified staff, a manifestation of the tendency to substitute capital for labour that has characterised industrial growth in Europe, Japan and North America.

Even if these technologies could be transferred on a large scale - which is unlikely because of bottlenecks in the production of energy equipment in the OECD countries - it could, in the final outcome go against the interest of the developing countries. Instead of stimulating the use of local resources for example, it would merely replace the Third World's present dependence on the Middle East for oil or capital by a far more permanent dependence on the developed countries for technological capability and capital.

The above considerations will be taken into account in the following summary review of the main energy sectors from the standpoint of the developing countries.

1. Hydrocarbons

As in the developed countries oil will obviously and necessarily remain one of the main sources of energy; one has only to think of its importance in agriculture. A considerable increase in oil and natural gas exploration programmes would therefore appear to be essential. It is, incidentally, probable that the developing countries possess large reserves that have not yet been discovered. On the other hand oil exploration needs time and money, to an extent which should not be underestimated particularly since the full capacity of the industries producing oil exploration equipment is already fully absorbed by the accelerated programmes in developed countries.

There can be no question of developing countries launching into their own R & D on oil and gas production technologies. The case would seem to be better met by transferring technologies already developed by the oil companies, although co-operation with the developed countries would be necessary in order to assist the developing countries towards a better grasp of the technologies involved. The areas where special effort might be made are the scientific disciplines and technologies involved in finding and evaluating gas and oil reserves so that the developing countries can improve their knowledge of their own energy resources and manage their exploitation.

A point in this connection is that the exploration programmes should not be confined to oil and natural gas deposits as such. Many Third World countries have considerable reserves of shale oil and it is probable that there are resources yet to be discovered in many other countries. Although the technologies for extracting oil from shale have not yet been perfected, this possibility merits careful study.

2. Coal

There are abundant reserves of coal in many Asian, African and Latin American countries. In many cases it could be easily mined by the open-cast method. At the moment, however, little is being done to exploit these resources and a special effort would therefore seem to be called for in this area particularly since, with the renewed interest of the developed countries in coal, the developing countries might well find a ready export market for it.

Mining itself is one of the technological fields where co-operation with the developed countries would appear to be necessary. On the other hand the mechanisation and automation characteristic of their technological development is unlikely to be the most suitable approach for the developing countries. At the moment, labour-intensive technologies would appear to be more suitable, provided safety and health questions are given the careful attention they call for.

Another area where research would be useful is the transport of coal by pipeline over the long distances that are often unavoidable in the developing countries. As regards the use of coal, gasification and liquefaction techniques are still too complex and costly to have any real interest for the developing countries; it would appear preferable to burn the coal as it is, either to generate electricity or for industrial or domestic purposes, provided environment problems are sufficiently taken into account.

3. Hydro-electric power

The hydro resources of the Third World are far from being as fully harnessed as those of the developed countries. Their exploitation, however, should be considered in conjunction with the problem of using water for agriculture - which could possibly be more advantageous. A further point is that little is yet known about the dynamics of nature and the possible effect of building large dams on the environment as a whole (effects on geological and geophysical equilibria, drawbacks as regards climatology, public health etc.). Experiments so far carried out suggest that caution is called for since the ecological equilibrium of many Third World countries, contrary to a widely held belief, is far less secure than that of the developed countries. It would therefore appear advisable to develop research along these lines.

4. Nuclear energy

This clearly involves dependence on the technological skills of the few countries that have a nuclear engineering capability. This dependence applies to all stages of the cycle - from the supply of fuel to the processing of wastes. None of the developing countries is currently in a position to undertake much responsibility even for part of the process.

Nuclear energy requires in any case very high capital investments and construction programmes now take even longer because of the saturated supply situation with regard to fuel and equipment on the world market. The dangers involved, too, are far from negligible particularly in countries not yet fully alive to the kind of problem involved. Lastly highly qualified manpower requirements are exceptionally critical.

The uses of nuclear energy would be restricted to a few special cases such as the generation of electricity for highly populated areas with the minimum high capacity that this implies, or sea water desalination.

5. Other sources of energy

Unlike certain types of energy, such as nuclear power (which is ill-suited to developing countries because of its complexity and high cost and the need for highly-qualified personnel), certain sources incorrectly described as "new" (solar, geothermal, etc.) are good examples of the forms of energy that might best correspond to certain conditions typical of many of these countries. It is through them, rather than the others, that a number of countries might achieve relative independence in energy and technology. They could also form an ideal field of application for a new approach to energy problems not drawing its immediate inspiration from the answers found in the industrialised countries.

One of the salient features of most of these sources is the fact that, by their nature or in some cases because of the present state of the art, they are more suited to local and restricted use than to large-scale and generalised application. For this reason, they would seem to be better tailored to the pattern of demand in many regions of developing countries where decentralisation and the low-power applications required would rule out the use of traditional power generation facilities. What is more, the range of applications they permit does not inevitably call for complex technology or excessively costly materials and is often feasible with the resources and technological level of the countries concerned. Three of these sources have an obvious interest: solar and geothermal energy and wind power.

a) Solar energy

Most of the developing countries have hot climates, and the sun is the most abundant source of energy they have although so far little use has been made of it. Because of the limited capital resources of very many of these countries and their lack of highly-qualified personnel, research on solar energy applications in such countries should concentrate on designing simple, economical and easily-operated plants. In other words, these solar energy applications should not be based on the sophisticated and costly systems under consideration for large-scale application in the developed countries but instead be centred on the use of low temperatures and flat collectors. Such simple systems are part of the R & D programmes of some of the OECD countries and should have significant applicability to the developing countries. In this way, solar energy could contribute to economic development through the local production of the various components (concentrators, piping, etc.). An outstanding example would be the use of solar energy for heating water; this could be made more general using collectors of a simple design that have already been proven in use.

Solar energy applications meeting many requirements common to all developing countries arise in many different areas: the life of the community, agriculture, craft trades and small-scale industry. At the present time, most of these countries have a vital need of water supplies and the use of solar energy to pump water from underground reserves is being given increasing attention. Current research is designed to widen the use of solar pumps for irrigation and to provide water to communities such as hospitals, welfare centres, etc. Solar energy development, however, should also cover other requirements such as small power units for agriculture and industry, kilns for baking bricks and other materials, water desalination, water heating, refrigeration (for food and medical products) and air-conditioning. Programmes are in hand for water-heating and air-conditioning plants but only on a small scale.

In the longer term, electricity generation using solar cells might be useful in developing countries because of its suitability for remote and isolated areas. Solar cells are still extremely costly, but if the efforts that are now being made to reduce their price are successful, this technology could well be within the reach of these countries and be used, among other things, for telecommunications (television).

Whilst the developing countries can and should endeavour to make use of solar energy themselves, close links will nevertheless be essential between them and the developed countries now engaged in research in this field. The conclusions of studies and analyses of the solar energy problem over several years have often spotlighted the fact that a major difficulty in drawing up an R & D programme for such countries is the dearth of knowledge about their energy requirements, their varying geographic and climatic conditions and, what is just as important, the psychological attitudes impeding certain developments. Research in this direction and the gaining of better basic knowledge so that solar energy potential might be evaluated in the light of the varying climates, insolation and position in relation to the sun, would be an excellent starting point for a co-operative research programme.

b) Geothermal energy

There are many developing countries with reserves of geothermal energy in Africa, Asia and Latin America. The present level of scientific knowledge is insufficient to enable these reserves to be evaluated with any accuracy but it is safe to say that certain of these countries have some of the biggest reserves in the world.

Applications which could be valuable to developing countries include heating and air-conditioning of houses and communities, and agricultural and even industrial processes, all of which could be powered from low-temperature sources, a technology already applied in certain industrialised countries. High-temperature sources producing dry steam could provide energy for electricity generation and be used in a relatively simple manner. The use of geothermal energy to generate electricity lends itself to a step-by-step approach with small and medium-power plants; it would therefore suit the requirements of developing countries and be more economical (in the case of dry steam) than other conventional sources.

Other forms of geothermal energy such as hot brines and dry rocks might have less interest in the present state of technology, because of the technical complexity, and high capital investment they involve. Nevertheless current pilot projects in this area should be evaluated with respect to potential applications to special needs.

For this source of energy, more than for any other, assistance from the developed countries should concentrate primarily on evaluating available sources, about which little is yet known. In this context, developing countries should be given the maximum of information and be enabled to visit the developed countries and attend training courses there.

c) Wind Power

Other sources restricted to local exploitation, such as wind power, could offer interesting possibilities. A highly important application of wind power is water pumping. This is by no means a new technology since Persians and Chinese were already using it 1700 years before the Christian era and it has allowed crops to be grown in the semi-arid areas of the south west of the United States, in Australia and in the south of Africa that could not have thrived in any other way.

The efforts that certain developed countries have been making for a number of years to use this source in developing countries have been concentrated on designing "wind turbine" systems which are completely different from the traditional windmill and far superior to it in many respects. In view of the relatively low cost involved in producing these systems for harnessing the force of the wind, they could well be used not only for pumping water but also for generating electricity in plants that would be small but nevertheless capable of producing the power required by communities, agriculture and small-scale industry.

The possibilities opened up by the research that is being done on these systems with a view to their application in the developing countries are also interesting for the developed countries where, because of the recent crisis, diversification in energy sources is again a topical issue. For this reason co-operation between these countries would appear likely to be established without any difficulty.

o

o o

Finally, another possible source of energy for developing countries is the utilisation of animal and agricultural waste. Production of synthetic fuels from waste is a research topic which at present interests developed countries and the results of pilot-plant investigations appear promising although the economics of the various processes available are still not competitive with conventional energy sources.

SUMMARY AND CONCLUSIONS

A. FUTURE PROSPECTS

This report has attempted to present a comprehensive view of the scientific and technological aspects of the energy problems, describing the R & D needs and opportunities in the OECD area. The main results, at least as far as the more promising and realistic solutions are concerned, can be summarized as follows:

The short-term (1974-1985)

In the short-term, OECD energy needs will be satisfied mainly by the same resources and technologies as exist today. Nevertheless, R & D activities, and more particularly those which already exist, can make significant contributions by discovering new resources and by improving existing technologies.

i) Oil and natural gas will continue to play a major role. Among the most important possible contributions of R & D are:
   - the improvement of prospection technologies on and off-shore;
   - the improvement of secondary and tertiary recovery methods;
   - better deep-drilling technologies;
   - with regard to off-shore technologies, it should be possible to produce oil in increasingly deep waters, up to an ocean depth of approximately 1,000 m.

ii) The direct use of coal for electricity generation will probably become more extensive. The most relevant contribution of R & D in this sector will be the development of new stack-gas cleaning technologies.

iii) Nuclear energy for electricity generation will be developed very rapidly. This development will essentially be based on already existing reactor types; mainly light water reactors and to a much smaller extent, heavy water reactors. In the nuclear energy sector, the greatest part of the R & D effort will have to focus on the fuel cycle.

Several energy sources could make contributions which, though limited in overall terms, might be quite significant in specific cases, for example:

- shale-oil in-situ extraction in the United States;
- tar-sand oil extraction in Canada;
- geothermal energy through dry steam;
- solar energy for space heating and cooling;
- synthetic fuels through pyrolysis of organic waste.

As to energy carriers, it is probable that electricity technologies will develop most rapidly. R & D will continuously improve the efficiency of electricity generation and reduce thermal pollution problems, through cooling towers, increasing utilisation of the residual heat of power plants, etc.

Finally, the analysis of energy systems should be pursued with greater speed and scientific effort. The contribution of systems analysis to a rational and efficient solution of energy problems could certainly make itself felt in the short-term, but it is also important for the medium and longer term.

The medium-term (1985-2000)

R & D now being pursued will have a much broader impact in the medium than in the short term. With regard to energy resources, it should be possible to produce oil from an ocean depth of considerably more than 1,000 m, and tertiary recovery methods will probably be applied on a broad, general scale. Better coal conversion (gasification, liquefaction, conversion into methanol), and advanced coal firing technologies will confirm the importance of coal for electricity generation and for synthetic fuel production. In addition, shale-oil extraction could spread to many countries which hitherto have made no systematic effort to discover and assess shale resources. Finally, nuclear energy will occupy an increasingly important place, following the introduction of fast-breeders and of high temperature gas cooled reactors.

Again, certain non-conventional energy sources could play a globally limited but locally very important role. Among these sources are geothermal energy through hot brines, solar energy for electricity generation, synthetic fuels through bio-conversion of plants or organic waste, wind-power etc.

Electricity will probably remain the energy carrier with the fastest development potential. Electricity generation could benefit from progress in advanced cycles and perhaps fuel cell technologies, electricity storage from progress such as that made with batteries and electricity transmission from that with underground cables. Moreover, electricity will profit from increasing possibilities of application such as electric cars. Other energy carriers will become increasingly important, for example methanol or others which can store and transport heat in the form of chemical bond energy.

It is in the medium-term that research to improve energy utilisation will be most successful. A more efficient energy utilisation at the level of individual technologies might help considerably to slow down the growth rate of energy demand.

The long-term (after 2000)

The long-term prospects are obviously much more uncertain. Fossil fuels will continue to play a noteworthy role, especially coal, as coal reserves are known to be sufficient for several centuries at the present rate of consumption. However, progressively, fossil fuels will have to give way to energy sources which have, if not "unlimited", then at least extremely abundant reserves. Thus, in the very long-term, fast breeders, controlled thermo-nuclear fusion, solar energy and last, perhaps geothermal energy through hot rocks will increasingly replace present energy sources.

As far as energy conservation is concerned, the biggest long-term impacts of R & D will among others, be found in the transportation and agricultural sectors.

B. POLICY CONCLUSIONS

From the above summary review, it follows clearly that the options and possibilities afforded by scientific and technical research are both very various and numerous. If these options and possibilities could be considered in isolation, the OECD countries could then look to the future with relative optimism. Many of the analyses and conclusions presented in this report support this view, precisely in those areas where energy problems appear to be reducible to matters of technology and capable of resolution in the form of research programmes.

Nevertheless, because of the incidence of political, economic and social factors, not to mention military ones, over-optimism is not justified. In spite of the multiplicity and diversity of medium and long-term prospects offered by science and technology both for the production of energy and its more rational use, caution is called for. Although energy among other things, provides an inexhaustible field for scientific and technical research, it is, in the first place, related to the structure and choices of society. Its future development will mainly depend on the political decisions of countries with regard to the nature of their economic growth and social structures.

Within the framework of these socio-political decisions, science and technology could make a vital contribution provided an immediate start is made on the vast R & D effort necessary to match the scale of the prospects already referred to. Moreover, to be fully effective, this R & D effort must be based on coherent policies at both national and international levels, covering the entirety of the field and taking full account of all the aspects involved.

In this connection it may be remarked that energy is a particularly good illustration of most of the problems currently facing science and technology policies(1). Satisfying continuous and changing energy requirements is a typical example of the "moving target" problem calling for sustained progress but which will never be resolved "once and for all". A further point, and this cannot be repeated often enough, is the need to take fully into account - at an early stage - the inherent time scale of all R & D activities. A research programme started today will take at least ten years, and more probably 15 or 20 years, to produce results.

Consequently, nothing could be less appropriate to the field of energy than a policy of "crash programmes" to respond to crises that have already taken place, framed against a rapidly-changing background of events and, in addition, perpetuating the detrimental effects of "stop-go" policies. Priorities should be established as part of a long-term strategy based on the evaluation of future problems and needs. Obviously, these evaluations, and therefore the relevant strategies, will need to be periodically revised in the light of new circumstances that will undoubtedly arise, provided these are not merely temporary phenomena but significant trend-changing factors. Moreover, since there is no single source of energy unquestionably leading the field for all time in terms of economic advantage, security of supply and environmental suitability, these strategies should be explicitly aimed at keeping open a sufficient number of alternative options. Above all, it is most important to be fully aware of the need to maintain a sense of consistency and continuity in implementing long-term R & D policies.

Energy also illustrates another principle in that R & D in this field must serve energy policies. It should not, however, be merely one of the tools of such policies; it should also identify those options that are technically feasible and indicate their various consequences. If it is to perform this function, R & D should not be confined to the strictly technical or economic aspects but should tackle the problem in its entirety, taking into account all its social, political and environmental implications.

The development and application of the studies on energy systems should provide an invaluable means of following this global approach. Furthermore, the implementation of energy R & D strategies as a function of energy policy should be formulated within the broad framework of the management of natural resources. In particular there are direct relationships between policies in support of energy R & D and materials R & D.

---

1) Cf. The Research System, Volume III, OECD, Paris 1975.

The study of the institutional aspects of the energy field also serves as an illustration of one of the permanent requirements (and problems) of science policy, namely, the need for co-ordination. For reasons that may be readily understood and to which reference has already been made, there are few fields featuring the same degree of dispersion and diversity as the structure of research in the various energy sectors. One of the most urgent - and most difficult - tasks of the bodies responsible for energy R & D policies is to co-ordinate the activities of the very many institutions concerned, since there is a tendency for the latter to preserve their autonomy. Political weight and financial resource are of importance for the co-ordinating body.

Energy R & D policy is to serve energy policy. To do so effectively it cannot be dissociated from overall scientific and technological policy. Apart from the effects that the increased priority accorded to energy is bound to have on other research sectors, energy research will in fact have to concern itself with questions of the mutual interaction between energy and social structures and the interdependence between it and other natural resources. This can only be achieved within the framework of science policy. For this reason, the report has repeatedly emphasized the fact that energy R & D cannot survive or flourish as a totally separate entity, but must be based on the overall advancement of science and technology in a large number of sectors and is, of course, dependent on the training of qualified engineers and scientists.

In view of the characteristic features of energy - its primary importance for the economic and social development of OECD countries, its international dimension, its increasing complexity, the key role of science and technology in this field and the size of R & D programmes involved - there is a vital need for developing international co-operation in science and technology. However, because of the very number and variety of the gaps or inadequacies pointed out above, there is no question of attempting to deal with all the sectors at once. Certain topics have to be selected where it is clear that there is an urgent need for greater co-operation between the OECD countries and where this can be achieved relatively quickly. This selection can be based on several criteria.

The quantitative or qualitative relevance of an R & D programme for energy production or conservation is certainly one of the main criteria. Another one is the absence of major obstacles, whether political, institutional or economic, that would prevent or hamper co-operation. This second criterion means in concrete terms that co-operation will often be easier in research than in development, and easier between government R & D programmes than between industrial R & D programmes where commercial interests are involved.

Another criterion is the existence of research programmes of sufficiently large a size in a number of countries, indicating broad interest for a certain subject.

Following these criteria at least fourteen subjects, as indicated below, can be considered as the most suitable for immediate international co-operation. These are:

1. Exploration and evaluation of energy resources
2. Underground or sea-bed storage of fossil fuels
3. The coal sector (mining, combustion, gasification and liquefaction techniques)
4. Safety of nuclear installations
5. Radioactive waste management
6. Thermonuclear fusion
7. Solar energy
8. Geothermal energy
9. Energy produced from industrial and urban wastes
10. Fuel cells
11. The use of heat as an energy carrier (covering not only waste heat from power stations but also the transportation and storage of heat)
12. New energy carriers (methanol, hydrogen and, more generally, carriers enabling energy to be transported in the form of the energy of the chemical bond)
13. Energy conservation (rational use of energy)
14. General studies on energy systems.

In addition it should be stressed that international co-operation in R & D must be viewed as an extension of national R & D and national efforts should be an integral part of overall energy R & D policy. Consequently _any joint action with regard to possible co-operation topics cannot be planned in isolation but must be fitted into the broader framework of international co-operation with respect to energy R & D policy_. Added to the necessary initiation of co-operative activities at least in the fields mentioned above, _there is therefore a need for the continuation of information exchange and analysis on national energy R & D policies_.

Finally, _in view of the seriousness of the energy problems of developing countries_, which affect not only them but also the developed countries, _it is important to undertake a thorough study of the ways in which science and technology can contribute to the solution of these problems_, and of the lines along which international co-operation in this field might be initiated and the conditions in which it might be organised.

ANNEXES

Annex I

MAIN ENERGY R & D PROGRAMMES

Annex I summarises the information provided by Member countries on their energy R & D activities. The main research programmes have been grouped following the classification adopted in Part II of the report "Scientific and Technological Prospects".

As only a few Member countries have provided usable data on the energy R & D effort carried out by industry, most of the R & D activities mentioned are carried out or supported by governments and nationalised industries. In view of the major role played by private firms in several energy sectors, it must therefore be realised that the following sections, although providing useful informations on the public effort, are far from giving a comprehensive view of all energy R & D activities in Member countries.

TABLE OF CONTENTS

1. RESOURCE ASSESSMENT AND PROSPECTION

Country	Financing	Institutions	Remarks
		a) Oil and Natural Gas	
Australia	Austr. $85,000 p.a.	CSIRO	Oil and gas exploration in the North-West Shelf area.
"	1973-74 Austr. $2.2 million	Bureau of Mineral Resources Geology and Geophysics, Department of Minerals and Energy.	Geological and Geophysical investigation of sedimentary basins on- and off-shore; Assessment of resources.
Austria			Oil and gas prospection.
Canada	1973-74: Can. $1,139,000 (oil and gas)	Responsibility and financing: Dept. of Energy, Mines and Resources, Ottawa.	Geological exploration of sedimentary basins on- and off-shore, determination of resources.
		Research Centres involved: Geolog. Survey of Canada Ottawa, Calgary, Dartmouth, Vancouver.	Research on petroleum geology and geochemistry, sedimentary petrology, etc.
Denmark		Industry	Oil and gas prospection in the North Sea, Jutland, Greenland.
France	1970: F 150 million 1971: F 141 million 1972: F 144 million (incl. R & D on advanced prospection, drilling and production technologies)	Institut français du pétrole and French oil industry (ERAP-SNPA, CFP) 143 mill.; foreign industry 1 mill. in 1972 with the collaboration of many R & D centres and administrations, e.g. CNEXO, ORSTOM, Institut de physique du globe, CGG.	Exploration and production research, mainly for off-shore oil, ocean-bottom mapping, development of geophysical instruments.
Ireland	"large amount"		Oil and gas prospection in the continental shelf off the south-west coast of Ireland.
Norway		Government Oil Directorate, NTNF.	Oil and gas prospection in the continental shelf.
Switzerland		Industry	Oil and natural gas prospection undertaken by a private firm. The "Office de l'économie énergétique" will follow the developments of this programme.
Turkey	Turkish L. 30 million for oil R & D Institute in planning stage, among others.	Ministry of Energy and Natural Resources and its Divisions, mainly Dept. of Petroleum Affairs and Mineral Exploration and Research Institute.	Oil and gas prospection, R & D on prospection, drilling, production methods.
United Kingdom	1972-73: $2.4 million(1) 1973-74: $3.6 million	The Institute of Geological Sciences (IGS) Some universities	Geological mapping of the United Kingdom, application of geophysics, geochemistry, geomagnetism and seismology to prospecting, development of instruments and techniques.
United States	FY 1974: $10.3 million (incl. geothermal sources, nuclear and other minerals, all fossil fuels). About $4 million are devoted to oil and gas; major increase anticipated for 1975, 4-6 times as large as present funds.	United States Geological Survey	Oil and gas prospection in selected on- and off-shore basins, studies on environmental effects, new exploration concepts, origin and migration of oil, petroleum stratigraphic maps, geochemical detection, bore hole geophysical exploration, detection of petroleum source rocks, continuous reading bore hole gravimeter.

1) Government expenditure covering also research in the Oil Sector (see Section 2a, on page 187) and in the Natural Gas Sector (see Section 3a, on page 189). Government expenditure is part of marine technology budget, which falls under the Ship and Marine Technology Requirements Board of the Department of Industry, and is also part of the Department of Energy's vote for petroleum production research. The work by Universities is funded by the Science Research Council.

1. RESOURCE ASSESSMENT AND PROSPECTION

Country	Financing	Institutions	Remarks
		b) Coal	
Australia	Austr. $55,000 p.a.	CSIRO	Analysis of assay data
"	1974-75: Austr. $500,000	Petroleum and Minerals Authority	Assessment of coal resources including drilling analysis.
"	1974-75: Austr. $70,000	Joint Coal Board	Assessment of coal resources of coking potential, singly and in blends.
Belgium	FB 5 million p.a.	Universities	Assessment of coal deposits and ores identification.
Canada	1973-74: Can. $155,000	Responsibility and financing: Dept. of Energy, Mines and Resources, Ottawa. Research Centres involved: a) Geological Survey of Canada, Ottawa and Calgary b) Mines Branch, Ottawa and Edmonton, Alberta	Assessment of Canada's coal resource potential, evaluation and classification of coals
Turkey		Ministry of Energy and Natural Resources; Mineral Exploration and Research Institute.	Exploration of coal and lignite, resource evaluation.
		Universities	Analysis of physical and chemical properties of the lignites in Turkey.
United Kingdom	1972-73: $240,000 1973-74: $240,000	The Institute of Geological Sciences (IGS); National Coal Board.	Covers R & D by both IGS and NCB on work similar to that done on assessment and prospection for oil and gas (see page) with techniques to suit coal prospecting.
United States	FY 1974: approx. $1 million, major increase anticipated for 1975.	United States Geological Survey	Mapping and appraisal of coal resources, studies of the geological factors relevant in exploration and mining, research on the origin and characteristics of low-sulphur coal, improved exploration technology, new system of coal data synthesis, computer processing and analysis will be initiated.
		c) Shale and Tar Sands (Oil Sands)	
Germany	1975: DM 0.5 million	Bundesanstalt für Bodenforschung (Federal) Office for Geological Research)	
Turkey		Cf. Section 1.b.	Exploration and evaluation of shales.
United States	FY 1974: less than $1 million	United States Geological Survey	Evaluation of shales, especially in Utah, Colorado, Wyoming. Studies of environmental and extraction problems. Evaluation of tar sands.

2. THE OIL SECTOR

Country	Financing	Institutions	Remarks
a) Prospection, Drilling, Production Technologies			
Australia	1973-74 Austr. $250,000	Bureau of Minerals, Geology and Geophysics, Department of Minerals and Energy.	Development of geophysical equipment and techniques. Reservoir geochemistry.
Canada	1973-1974: Can. $4,773,000	Department of Energy, Mines and Resources, (Mines Branch), Ottawa. Department of Industry, Trade and Commerce.	Offshore petroleum production systems; steam cracking naptha from crude oil; pipeline materials evaluation.
France	Cf. Section 1, a (page 185).	Cf. Section 1, a (page 185).	Development of floating, dynamically anchored drilling-rigs (ship "Pelican") geophysical equipment, Arctic zone oil equipment, exploration R & D in deep seas, development of drilling, well-head and production equipment up to an ocean depth of 1,000 m. by 1980, and, later, up to 2,500/3,000 m. under water, laying and repair of deep-sea-oil-pipelines.
Germany	1974: DM14 million 1975: DM30 million 1976: DM52 million 1977: DM78 million	Organised and financed for approx. two-thirds, by the Federal Government.	Improved exploration for oil and gas by better seismic and geochemical methods, emphasis on deep-drilling technologies, and on off-shore technologies.
Japan	1970-1974: approx. Yen 5 billion	Ministry of International Trade and Industry	Development of remote-controlled undersea oil-drilling rigs (automatic digging machine linked to surface power supply device, other devices for mud water circulation and cementing, transportation and communication).
Netherlands		TNO; Industry (Shell).	Development of drilling platforms. Improving exploration (mainly seismic) techniques, extraction and off-shore engineering technologies.
Norway	1973: N.Kr.24.4 million, of which: 11 million from Governmental Oil Directorate, 13.4 million from NTNF	Government, NTNF.	Deep seismic exploration technologies, geological and bottom mapping, oceanography, pollution problems, marine technology and instrumentation.
United Kingdom	(See amounts and footnote 1 on page)	Department of Industry, Department of Energy.	Marine technologies.
United States	FY 1974: $3,393,000 (incl. gas, heavy oils, tar sands)	United States Bureau of Mines	Secondary and tertiary recovery of oil, reservoir geochemistry.
b) Transportation			
Canada	FY 1973-1974: Can. $2,350,000	Canadian Transport Commission, Ministry of Transport, National Research Council, Department of Energy, Mines & Resources.	Arctic railway studies; slurry coal pipelining; oil and gas pipeline technology.

2. THE OIL SECTOR

Country	Financing	Institutions	Remarks
b) Transportation (Cont'd)			
Netherlands		Industry (Shell)	Oil-transportation from off-shore well-heads.
United States	Proposed for 10-year period: $400 million	Maritime Administration. Financing: - industry: 60 per cent - government: 40 per cent.	Arctic marine transportation R & D. Icebreaking, submarine tankers, ports, terminals, logistics requirement.
	Proposed for 10-year period: $300 million	Maritime Administration, Industry and Government financing.	Marine transportation of oil and slurried coal between East Coast and Alaska or Japan via Panama.
c) Storage			
France		Atomic Energy Commission	R & D to create underground cavities through nuclear explosions, for the storage of liquid fuels.
Japan	FY 1973: Yen 67 million FY 1974: Yen 79 million	Ministry of International Trade and Industry	Storage of oil on the sea bed, in large-scale vessels.
Sweden	1973-1974: Sw. Kr. 250,000	FOA, FMV, OEF (Dept. of Defense).	Storage stability of fuel oil.
Sweden	1973-1974: Sw. Kr. 120,000	FOA, FortF, STU, and others.	Corrosion in fuel oil tanks; development of corrosion inhibitors.
Switzerland	1974-1975: FS 6 million (3 million p.a.)	Industry; Office de l'économie énergétique	Study of the possibilities of storing oil and gas in Swiss geological formations. Seismic surveys and drilling.

3. NATURAL GAS

Country	Financing	Institutions	Remarks
	a) Production, Conversion into LNG, Utilisation		
France		Gaz de France	Studies to make gas interchangeable in domestic, commercial and industrial equipment.
United Kingdom	(See amounts and footnote 1 on page 185).		
United States	FY 1972: $6,800,000 FY 1973: $5,868,000 FY 1974: $ 950,000	Atomic Energy Commission	Recovering gas from tight, impermeable rocks by specially designed nuclear explosives.
" "	Part of $3,393,000	Bureau of Mines	Hydraulic and chemical explosive fracturing to recover gas.
" "	Approx. $850,000 p.a. (Total cost of programme amounts to approx. $10 million)	National Bureau of Standards	Development of liquefied natural gas technologies, refrigeration system for transportation.
	b) Transportation as Gas and as LNG		
Canada	1970-1973: approx. Can. $30 million	Arctic Gas (28 industrial companies)	R & D on engineering economics and environmental impact of a pipeline system across the Arctic and Sub-Arctic, in permafrost conditions.
"	Proposed for the next years: Can. $20 million	Polar Gas Project (5 industrial companies)	R & D on underwater pipelines between the Arctic Islands, at depth up to 300 m. Marine surveys, ecological studies. Other concepts studied; ice breaker LNG tankers, giant aircraft, electricity transmission.
France		Gaz de France and two customers	Materials R & D for cryogenic hulls in LNG ships.
Germany	DM 1 million		Storage and transport of LNG.
United Kingdom		British Gas Corporation	R & D on pipes, components, fracture behaviour of large pipelines, stress corrosion, inspection devices in high pressure systems.
United States	Proposed for 3 years: $5 million	Maritime Administration; industry-government financing (50/50 per cent).	R & D on construction of United States flag LNG carriers.
	c) Storage		
Austria			Underground storage of gas.
France		Gaz de France, R & D carried out in Ecole des Mines, Paris.	R & D on rock mechanics and cavity stability in salt, nuclear and natural cavities.

3. NATURAL GAS

Country	Financing	Institutions	Remarks
	c) Storage (Cont'd)		
France	Approx. F 1 million p.a.	Gaz de France and ELF-ERAP.	R & D on storage of gas in aquifers.
"	Proposed for the next years: approx. F 5 to 8 million p.a.	Gaz de France and ELF-ERAP, Geostock.	R & D on storage of gas in Karst-cavities.
"	Proposed for 1974-78: F 5 to 10 million (incl. oil storage; see Section 2, c, page 188).	Gaz de France, Atomic Energy Commission.	R & D on storage of gas in cavities created by nuclear explosion near to sea coasts.
Switzerland	1974-75: FS 6 million (3 million p.a.)	Industry; Office de l'économie énergétique.	Study of the possibilities of storing oil and gas in Swiss geological formations. Seismic surveys and drilling.
United Kingdom	1972-73: $480,000 1973-74: $480,000	British Gas Corporation	Economics and safety of high pressure storage in pressure vessels and pipelines, salt cavern storage, LNG storage.

4. SHALE, TAR SANDS, HEAVY OIL

Country	Financing	Institutions	Remarks
Canada	approx. Can. $750,000	Dept. of Energy, Mines and Resources, Ottawa. Research Centers involved: a) Mines Branch, Ottawa b) Geological Survey of Canada.	R & D on purifying and upgrading heavy oils and tar sands; underground recovery of oil from tar sands.
Canada (Province of Alberta)	5-year programme of Can. $100 million	Gov. of Alberta, Alberta Oil Sands Technology and Research Authority.	R & D on in-situ production of oil from deep tar sands. Controlled underground heat generation.
France	1974: Frs. 2.1 million	IFP/Société pétrolière française/BRGM/Cerchar (DGRST).	Evaluation of national resources.
United States	part of FY 1974: $3,393,000	Bureau of Mines, AEC.	Heavy oils and tar sand recovery R & D.
" "	FY 1974: $ 987,000	" " "	Characteristics of oil shale.
" "	FY 1974: $1,385,000	" " "	In situ retorting of oil shale. Field fracturing and recovery R & D, ground water studies.
" "	FY 1974: $ 100,000 expected to expand 40-60 times	" " "	Oil shales mining technologies, waste disposal, land reclamation R & D, water problems.
" "	FY 1973: $ 160,000 FY 1974: $1,075,000	Atomic Energy Commission	In-situ oil shale retorting; R & D on shale permeability and nuclear fracturing of oil shale rocks.

5. THE COAL SECTOR

Country	Financing	Institutions	Remarks
		a) Mining	
Australia		Coal mining and beneficiation sub-committee of the National Coal Research Advisory Committee.	R & D on cost reductions of mining.
"	1974-75: Austr. $80,000	Australian Coal Industries Laboratories	Research on mining thick seams (CRETA measures) at depth.
"	1972-74: Austr. $100,000 1974-77: Austr. $ 90,000 planned	Sponsors: 3 coal producers; R & D carried out by Australian Coal Research Laboratories.	Underground strata control research.
Belgium	FB 50 million p.a.	Universities and Institut national des industries extractives.	Study of improvement of mining techniques and more particularly of automation.
Canada	1972-73: Can. $167,000 (part of coal programme of Can. $1,513,000).	Dept. of Energy, Mines and Resources. Research Centres involved: a) Mines Branch, Ottawa and Edmonton, Alberta b) Geological Survey of Canada, Ottawa and Calgary.	R & D on underground mining technologies, especially on new technologies for the thick deep seams containing a large part of Canada's western coal. R & D on gas outbursts and on roof stability in coal mines.
"		A coal producer in British Columbia and a Japanese mining company.	R & D on a hydro-mechanical system to extract coal from a 50 foot coal seam, having a pitch averaging between 30 and 50 degrees. High pressure water blasts the coal from the seam.
France	1973: F 50 million	Financed by Charbonnages de France, local mining firms and public funds; research carried out by Cerchar and coal mines.	General mining problems such as control of seams, pressure, dust, ventilation, gas outburst, fire hazards, telecommunication, logistics (47 per cent); coal mining technologies (21 per cent); transportation technologies (11 per cent).
Germany	1974: DM 102 million 1975: DM 114 million 1976: DM 113 million 1977: DM 109 million	Steinkohlenbergbauverein (Bituminous Coal Mining Association); industrial share 50 per cent.	Safety programme of the Land Nordrhein-Westfalen; current projects of Steinkohlenbergbauverein, partly funded by ECSC.
Ireland	£10,000 p.a. for 2 years	Sponsor: Department of Industry and Commerce. Research carried out by industry.	Development of a high-ash coal block for mechanical extraction in the Arigna coal fields, to be used for electricity generation nearby.
United Kingdom	1972-73: $6,200,000 1973-74: $7,050,000	National Coal Board	R & D on increasing efficiency of deep coal mining, mainly mechanised longwall face mining. Improvements of coal cutting machinery, of strata control techniques, mechanisation of face end operations, improved environmental, communications and transport systems, prototype machinery testing, better coal preparation.
" "	1972-73: $3,120,000 1973-74: $3,580,000	Safety and Mines Research Establishment (SMRE)	Better mine safety, prevention of accidents from gas, fire, dust.
United States	FY 1974: $1.05 million	Bureau of Mines	Surface coal mining (integrated extraction and reclamation systems, new equipment to increase production, improve overburden handling, surface and ground-water studies, land stabilization and revitalisation).

5. THE COAL SECTOR

Country	Financing	Institutions	Remarks
		a) Mining (Cont'd)	
United States	FY 1974: $6.45 million	Bureau of Mines	Underground coal mining (increasing productivity from 12 tons per shift to 30 tons by 1985. Automation and remote control, haulage systems, roof support, high speed development of mines, methane recovery methods etc.).
" "	FY 1974: $3.3 million	" "	Health engineering (dust abatement, noise research, toxic and noxious gases).
" "	FY 1974: $21 million	" "	Safety engineering (ground pressure, roof support, fire and explosion prevention, work accidents, methane control, systems engineering to design safety into mining operations, post disaster survival and rescue research).
	b) Improved Combustion, Pollution Abatement		
Australia		National Coal Research Advisory Committee, Sub-Committee on Combustion and Gasification of Coal and Coke.	Improved combustion R & D (particulate matter emission, drying of brown coal, spouted bed coal technology, nitrogen oxides in combustion, fluidised bed carbonisation, coal pulverising, industrial boiler adaptation to clean air legislation, flame radiation).
"	Austr. $195,000	CSIRO	Flyash and nitrogen oxides formation research to reduce air pollution.
"		"	Disposal of coal washeries waste, or conversion into energy.
"	1975: Austr. $125,000	CSIRO and Joint Coal Board	Fluidised bed combustion of coal preparation plant refuse.
"	1974-75: Austr. $50,000	Australian Coal Industry Research Laboratories.	Mine and coal preparation plant effluent control.
Canada	1972-73: Can. $885,000 (incl. land disturbance through strip-mining, and underground mining pollution).	Department of Energy, Mines and Resources. Research Centre involved: Mines Branch, Ottawa.	Air pollution control (particulate matter, sulphur dioxide); water pollution control (coal washeries); reduction of pollutants in mines.
Germany	$3 million p.a.	Bergbauforschung (Mining Research)	Development of new hard-coal power stations with fluidised bed combustion.
"	DM 24 million p.a. (of which DM 9.5 million supplied by industry)		Current projects of Steinkohlenbergbauverein (Bituminous Coal Mining Association)
Netherlands		Industry (Shell)	Stack-gas desulphurisation R & D.
United Kingdom	(1)	Central Electricity Generating Board (CEGB)	R & D to support and improve existing fossil fuelled electricity generating designs.

1) Expenditure for R & D in this field is included under the heading "Electricity Generation, improvement of power plants", Section IV, 1a, (ii) on page 220. A detailed breakdown is not available.

5. THE COAL SECTOR

Country	Financing	Institutions	Remarks
	b) Improved Combustion, Pollution Abatement (Cont'd)		
United Kingdom	1972-73: $303,000 1973-74: $405,000	Combustion Systems Ltd. (joint company of NRDC, N.C.B. & B.P.)	Fluidised bed combustion R & D (atmospheric pressure combustion, low sulphur oxide emission), pressurised combustion for combined power cycles, de-watering of colliery tailings.
" "	1972-73: $720,000 1973-74: p.a. for each year	British Gas Corporation	R & D to improve efficiency of gas appliances and for industrial purposes.
United States	Approximate funding levels for improved combustion processes programme, as follows:	Department of the Interior	
" "	$15 million	" "	Fluidised bed (atmospheric pressure) pilot plant scale boiler (30 MW) at West Virginia - initial operation 1975.
" "	$55 million	" "	Fluidised bed (pressurised) pilot scale (30 MW to 70 MW) boiler based on British and Argonne Lab. results. Start up 1977-78.
" "	$35 million	" "	Fluidised bed (pressurised with excess air) pilot scale boiler with direct power extraction. 25 MW. Start up 1978-79.
" "	$40 million	" "	Pressure boiler fluidised bed pilot scale, suited to advanced power cycles. Start up 1979.
" "	$150 million	" "	Fluidised bed boiler (atmospheric) demonstration/commercial scale. Multihundred MW size based on 30 MW pilot size. Start up 1978-79.
" "	$175 million	" "	Fluidised bed boiler (pressure) demonstration/commercial scale. Multihundred MW size based on pilot size. Start up 1979-80.
" "	$10 million	" "	Fluidised bed engineering study. Fluidisation feeding.
" "	FY 1974: $1.59 million (incl. strip-mining damage)	Bureau of Mines (Department of the Interior)	R & D to manage coal wastes and pollutants (mined-land reclamation, stack-gas desulphurisation, coal desulphurisation, toxic elements in coals and gases).
" "	FY 1974: $100,000	Bureau of Mines (Department of the Interior)	SO_2 Gas cleaning system for coal fired power plant, citrate process.
" "	FY 1974: $1,715,000 (incl. MHD carbonisation, coal drying and storage)	" "	Corrosion in coal power plant boilers, combustion of process chars.
	c) Coal Conversion into Gas, Low-BTU and High-BTU		
Canada, Province of Alberta		Alberta Research Council	R & D on underground gasification for Low-BTU gas, first field trial 1976. Progress would also benefit in situ tar sand extraction technologies.

5. THE COAL SECTOR

c) Coal Conversion into Gas, Low-BTU and High-BTU (Cont'd)

Country	Financing	Institutions	Remarks
Canada Province of Ontario		Ontario Research Foundation	R & D on producing High-BTU gas from James Bay lignite deposits.
Germany	1974: DM.95 million 1975: DM.99 million 1976: DM.106 million 1977: DM.108 million	Approx. 2/3 financed by the Federal Government. R & D is carried out by Bergbauforschung, Essen, and industry, e.g. STEAG AG, Essen, Lurgi Corporation and the Aachen College of Technology.	Most interesting are the projects for pressurised gasification combined with a gas and a steam turbine (conversion factor from thermal to electric energy up to 45 per cent), and the application of nuclear process heat from HTGR's to coal gasification. Other projects are: fixed-bed gasification under pressure, fluidised-bed carbonisation, pulverised coal gasification under pressure, tube furnace gasification of lignite, using its natural moisture, high-temperature Winkler gasification of lignite, fluidised-dust gasification for gas/steam turbines, pressure gasification of lignite, slurried with oil.
"	DM.24 million p.a. (of which DM.9.5 million supplied by industry)		Current projects of Steinkohlenbergbauverein (Bituminous Coal Mining Association).
Japan	FY 1974: Yen 435 million	Ministry of International Trade and Industry	"Sunshine Project". Several coal gasification (partial oxidation, hydrogenation, plasma gasification) and several liquefaction processes have been proposed and will be tested or developed. Programmes forseen up to the year 2000.
Netherlands		Industry (Shell)	Gasification of coal and heavy fuel oil.
United Kingdom		National Coal Board	Preparation for low-BTU gas: R & D on fluidised bed gasification employing air/steam as gasifying medium. Part of the NCB component of the ECSC programme for 1974.
"	1972-73: $ 240,000 1973-74: $1,680,000	NRDC, NCB and BCURA under contract from COGAS Development Co. United States.	High-BTU gas: COGAS - pilot plant for gasification of char from COED process.
"	(Not available)	B.G.C. (Westfield plant)	High-BTU gas: Lurgi - Catalytic conversion of Lurgi gas to methane in commercial scale unit Lurgi - pilot plant on behaviour of different American coals gasification. Funded in the United States.
"		B.G.C. (Westfield plant)	Preparation for development of full-scale slagging gasification technology, commencing in 74/75. Being funded by the United States.

5. THE COAL SECTOR

Country	Financing	Institutions	Remarks
c) Coal Conversion into Gas, Low-BTU and High-BTU (Cont'd)			
United States	FY 1974: $3.84 million (incl. liquefaction R & D)	Bureau of Mines	Low- and High-BTU - gasification, underground gasification of coal, liquefaction, coal conversion to hydrogen.
" "	FY 1974: $800,000 (incl. liquefaction R & D)	Sponsor: National Science Foundation. R & D carried out by several universities.	Fundamentals of chemical/physical processes and properties of coal, improved gas/solid/liquid separation systems, catalyst development, improved materials and mechanical systems, innovative approaches to coal conversion, coal systems evaluation research.
" "	Approximate funding requirements:	Department of the Interior	<u>Low-BTU-gasification</u>:
	1) $120 million	"	Entrained-bed gasifier (pressure) - pilot plant (30 MWe-50 MWe). 1977 initial operation.
	2) $120 million	"	Fluidised-bed gasifier (pressure) - pilot plant (30 MWe-50 MWe). 1978.
	3) $ 25 million	"	Entrained-bed gasifier (atmospheric pressure) - pilot scale operation. 1975-79.
	4) $ 15 million	"	Fluidised-bed gasifier (atmospheric pressure), PEDU scale. 1976.
	5) $ 40 million	"	Molten salt gasifier, pilot scale. 1976.
	6) $ 50 million	"	Fluidised/fixed bed stirred gasifier, both pressure and atmospheric capability. 1976.
	7) $ 45 million	"	Fluidised bed gasifier, pressurised and hydrogasification, 1976.
	8) $ 15 million	"	Suspension bed gasifier, low pressure, 1978.
	9) $ 40 million	"	Slurry firing with clean up gasifier, 1976.
	10) $ 55 million	"	Agglomerating bed process, 1978.
	11) $ 30 million	"	High-temperature gas clean up, using molten salt, 1977.
	12) $ 25 million	"	High-temperature gas clean up, using hot dolomite, 1979.
	13) $ 15 million	"	Stirred-bed gasifier.
			<u>High-BTU-gasification</u>:
	1) $10 million to complete	"	HYGAS
	2) $35 million	"	BI-GAS
	3) $15 million to complete	"	CO_2 Acceptor-fluid bed (lignite)
	4) $95 million	"	SYNTHANE (SNG from all coal types)
	5) $10 million	"	Liquid phase methanation (nickel catalyst)

5. THE COAL SECTOR

Country	Financing	Institutions	Remarks
	c) Coal Conversion into Gas, Low-BTU and High-BTU (Cont'd)		
United States	6) $20 million	Department of the Interior	Steam-Iron (uses char from HYGAS)
	7) $5 million	"	Self-Agglomerating (synthesis gas)
	8) $15 million	"	Materials development for gasification plants.
	9) $15 million	"	Mechanical equipment development
	10) $20 million	"	Fundamental process development
	11) $400 million	"	Demonstration plant for SNG from coal.
	d) Coal Liquefaction		
Australia	1974-75: Austr. $135,000	Australian Coal Industry Research Laboratories	Survey on solvent refined coal.
Germany	1974: DM.48 million 1975: DM.51 million 1976: DM.54 million 1977: DM.55 million	Cf. Section 5, c, page 195.	R & D focusses on production of fuel oil from coal, rather than gasoline. Hydrogenation of coal to produce heavy heating oil which is environmentally clean. Additional hydrogenation of coal oil is studied to produce middle distillates (heating and diesel oil).
"	DM.24 million p.a. (of which 9.5 millions supplied by industry)		Current projects of Steinkohlenbergbauverein (Bituminous Coal Mining Association).
Japan		Cf. Section 5, c, page 195.	Cf. Section 5, c, page 195.
United Kingdom	1972-73: $451,000 1973-74: $497,000	National Coal Board	R & D on extraction of liquids from coal with component gases close to their critical temperature (lab. scale) and on extraction with a coal based liquid solvent (small pilot plant scale). Part of the NCB component of the ECSC programme.
United States	Included in Gasification R & D (Cf. Section 5, c, page 196.	Bureau of Mines	Cf. Section 5, c, page 196.
" "	Included in coal R & D (Cf. Section 5, c, page 196.	National Science Foundation	Cf. Section 5, c, page 196.
" "	Approximate funding requirements:	Department of the Interior	
	1) $45 million	"	Completion of Solvent Refined Coal pilot plant.
	2) $100 million	"	Direct hydrogenation (H-Coal Process) pilot plant, ebullated bed.
	3) $70 million	"	Combination project (carbonisation, hydrogenation, desulphurisation) to produce char and coke. Process plant.

5. THE COAL SECTOR

Country	Financing	Institutions	Remarks
		d) Coal Liquefaction (Cont'd)	
United States	4) $40 million	Department of the Interior	Testing of multiple processes, alternate hydrogenation systems.
	5) $100 million	"	Methanol and other liquids from coal, pilot plant tests.
	6) $175 million	"	Supporting projects to seek other, better or cheaper processes (e.g. nuclear heat application to coal liquefaction)
	7) $125 million	"	Demonstration plant designs and evaluations.
		e) Peat	
Finland	Approx. Fmk. 140,000 p.a. for 3 years	Sponsor: Ministry of Trade and Industry. R & D carried out in industry and EKONO Oy.	R & D on promoting and using peat for heat generation. Peat burning studies in different boiler types.
Ireland	1971: £145,500 1972: £122,500 1973: £130,000 1974: £145,000 1975: £160,000 1976: £175,000	Sponsor: The Irish Peat Board (semi-governmental body).	Survey of Irish peat deposits, improvements of plant and production techniques for peat, development of new installations for peat loading and packaging, production of new peat-anthracite smokeless fuel.
Sweden		Board of Economic Defense (ÖEF)	Peat for heating and other purposes.
f) Other Problems: Coal Preparation, Coking, Transportation, etc. (Many countries have included the items mentioned in this section into "Mining", "Combustion" or "Conversion")			
Australia		National Coal Research Advisory Committee	R & D on coal oxidation, which affects coking properties and can lead to spontaneous combustion.
"		Broken Hill Proprietry Ltd.	Formed coke pilot plant operations.
"	1974-75: Austr. $10,000	Australian Coal Industry Research Laboratories.	Formed coke survey.
Canada	1972-73: Part of coal R & D programme of Can. $1,513,000	Dept. of Energy, Mines and Resources. Research Centres involved: a) Mines Branch, Ottawa and Edmonton, Alberta b) Geological Survey of Canada, Ottawa, Calgary.	R & D on beneficiation of different coals, improvement of conventional and new coking technologies (hot briquetting), R & D on economic desulphurisation of Cape Breton coking coal, integrated coal washing circuits.
"		Dept. of Energy, Mines and Resources	Investigations of coal transportation systems (coal-water-slurry or coal-oil-slurry pipelines, unit train concepts for large scale coal movements).

5. THE COAL SECTOR

Country	Financing	Institutions	Remarks
f) Other Problems: Coal Preparation, Coking, Transportation, etc. (Many countries have included the items mentioned in this section into "Mining", "Combustion" or "Conversion") (Cont'd)			
France	Included in "mining" (Cf. section 5, a, page 192).	Cf. section 5, a, page 192.	Coal preparation and coal coking R & D. Improvement of conventional and development of new coking technologies (Coaltek) with reduced pollution effects.
Germany	DM.24 million p.a. (of which DM.9.5 million supplied by industry)		Current projects of Steinkohlenbergbauverein (Bituminous Coal Mining Association).
"	1974: $1 million	Bergbauforschung	Formed coke
Netherlands		Industry (DSM)	R & D on coal-processing and other coal technologies.
United Kingdom	1972-73: $2,900,000 1973-74: $3,360,000	National Coal Board	Coal preparation and transportation included under "Mining" (section 5, a, page 192). Expenditure shown covers only coking. R & D by N.B.C. It includes contribution by N.C.B. to ECSC.

1. URANIUM RESOURCES - EXPLORATION, ASSESSMENT AND EXPLOITATION

Country	Financing	Institutions	Remarks
Austria			Uranium prospection.
Australia	Australian $40,000	CSIRO	Exploration of uranium-resources.
"	Austr. $650,000	Bureau of Mineral Resources, Geology and Geophysics, Department of Minerals and Energy.	Study of geological controls and genesis of uranium. Assessment of resources.
Canada	1973-74: Can. $862,000	EMR (Department of Energy, Mines and Resources); Geological Survey of Canada (Mines Branch).	Exploration for new uranium deposits along with other projects on the geology of uranium, geochemistry of radium and in situ leaching of uranium from an orebody.
"	1972-73: $548,000	Eldorado Nuclear Ltd.	Recovery of uranium from ores; refined uranium-products process research.
Denmark		Private firms	Uranium prospection
Finland	Fmk. 1-1.5 million p.a.	Main financing by the Ministry of Trade and Industry with the participation of two State-owned mining companies. Programme mainly carried out by the Geological Survey of Finland with the two State-owned companies involved in the field work.	Uranium prospection.
France		CEA (Commissariat à l'énergie atomique)	Uranium prospection and mining research.
Germany	DM.28 million	Activities performed by two industrial firms and funded up to 75 per cent by Federal grants.	Uranium prospection.
Ireland		Dept. of Chemistry, University College, Galway.	Extraction of uranium from sea water by ion exchange methods.
Japan	1972: Yen 145 million 1973: Yen 179 million 1974: Yen 303 million	Power Reactor and Nuclear Development Corporation (PNC).	Overseas survey of nuclear fuel.
Sweden	S.Kr. 20,000 p.a.	Dept. of Metallurgy, KTH (Royal Institute of Technology)	Biogene soaking of uranium shales.
"	FY 1973-74: S.Kr. 2.6 million	AB Atomenergi in collaboration with the Atomic Energy Company and the State Power Board.	Production of uranium from shale.
"		National Geological Survey Institute.	Uranium prospection.
Switzerland	Sw.Frs. 100,000 p.a.	Division de la science et de la recherce	Prospection and assessment of uranium resources.
United Kingdom		BNFL (British Nuclear Fuels Limited)	Assessment of uranium resources.
"	1972-73: $240,000 1973-74: $240,000	UKAEA	Exploration of United Kingdom for deposits of uranium.
United States	FY 1974: $1 million	USGS (Geological Survey)	Assessment of radioactive mineral resources.
"	FY 1974: $200,000	Bureau of Mines	Development of ion exchange and other processes for extracting uranium from dilute solutions.

2. FUEL CYCLE

Country	Financing	Institutions	Remarks
		a) Uranium Enrichment	
Australia		Atomic Energy Commission	Ultra centrifuge process.
Belgium	FB 7,500 million for a period of 5 to 7 years	Participation in the international EURODIF programme	Gazeous diffusion
France		Participation in the international EURODIF programme.	Gazeous diffusion and ultra-centrifuge process.
Germany	DM.114 million	Collaboration with the United Kingdom and the Netherlands through the international firms CENTEC and URENCO	Ultra centrifuge process.
Japan	1972: Yen 1,402 million 1973: Yen 5,241 million 1974: Yen 9,591 million	Power Reactor and Nuclear Fuel Development Corporation	Centrifuge process.
"	1972: Yen 521 million 1973: Yen 227 million 1974: Yen 208 million	Japan Atomic Energy Research Institute (JAERI)	Gazeous diffusion process.
Netherlands		UEN. Collaboration with Germany and the United Kingdom as mentioned above.	Ultra centrifuge process.
"		Universities	Isotope separation by fast rotating plasmas.
Sweden		AB Atomenergi	Gaseous diffusion.
"		"	Minor effort on gas centrifugation.
United Kingdom	1972-73: $5,520,000 1973-74: $10,100,000	BNFL CENTEC URENCO	R & D on centrifuge. Collaboration with Germany and Holland in ultra centrifuge process through CENTEC and URENCO. Figures shown for 72-73 are for BNFL only but those for 73-74 include contributions of $8,900,000 to CENTEC and URENCO.
		b) Other	
Belgium	BF 6.5 million p.a.	Centre d'études nucléaires (CEN), MOL; Universities.	Fuel Reprocessing.
France		CEA	Fuel reprocessing.
Germany	DM.70 million	KEWA, EUROCHEMIC, URG.	Fuel reprocessing, Plutonium recycling.
Japan	1972: Yen 1,725 million 1973: Yen 1,260 million 1974: Yen 1,059 million	PNC (Power Reactor and Nuclear Fuel Development Corporation).	Fuel reprocessing.
Netherlands		Interfuel	Prototype fuel element fabrication.
Switzerland		Federal administration	Fuel reprocessing.
"	SF 2 million p.a.	Institut fédéral de recherches en matière de réacteurs	Fabrication of fuel elements, such as small vibrating balls, composed of a mixture of plutonium and uranium carbides. Plutonium content: 10-20 per cent. Programme starting date: 1970.

2. FUEL CYCLE

Country	Financing	Institutions	Remarks
		b) Other (Cont'd)	
United Kingdom	(Not Available)	BNFC UKAEA	Depleted uranium used in fast reactors and also in centrifuge and consequent expenditure included under those subheads. Plutonium used for thermal fast reactors. Waste charged to "Radio-active waste management". R & D on reactor elements charged to reactors' subheads.
United States	FY 1973: $50.3 million FY 1974: $57.3 million FY 1975: $66.0 million	AEC	Fuel reprocessing.

3. LIGHT WATER REACTORS

Country	Financing	Institutions	Remarks
Belgium	BF 165 million p.a.	Centre d'études nucléaires (CEN), MOL; Industry.	Improvement of equipment and fuel.
Finland	Fmk 600,000 p.a.	Research financed by the Ministry of Trade and Industry, performed by the Technical Research Centre of Finland.	Development of a computer programme system for the fuel management in LWRs.
France	Frs. 100 million p.a. for 3-4 years	CEA in collaboration with EDF (Electricité de France)	Improvement of PWRs as well as BWRs.
Germany	Government financing: 1973: DM.30 million 1974: DM.23 million 1975: DM.22 million 1976: DM.21 million	R & D work mainly performed and financed by industry: Kraftwerk Union and Brown Boveri Reaktorbau. Government contribution and research concentrated on fuel elements and reactor safety.	
Japan	1972: Yen 262 million 1973: Yen 396 million 1974: Yen 348 million	PNC, JAERI.	
Netherlands		RCN	
Norway	N.Kr.20 million per annum. (1/3 financed by the Norwegian Government, 2/3 by foreign signatories to the Halden project)	Institutt for Atomenergi. Work at Halden operated as an OECD project.	Experimental fuel performance studies. Computer based control and supervision systems.
"	N.Kr.3.5 million p.a.	Institutt for Atomenergi.	Computer programme system for fuel management. Development of zircaloy alloys and canning tubes.
Sweden	FY 1973-74: S.Kr.25 million	AB Atomenergi	Among the various projects involved, three have unique features: 50MW R2 materials testing reactor; zero power high temperature reactor KRITZ; prestressed concrete pressure vessels for LWR.
"		ASEA-ATOM	Improvement of ASEA-ATOM Boiling Water Reactors.
"		State Power Board	Computer programmes for the calculation of power dissipation and burnout, and of void and temperature distribution in LWRs; hazard analysis in PWR and BWR stations: water chemistry and corrosion in LWRs.
United Kingdom	1972-73: 1973-74: Negligible	UKAEA	Expenditure on this is minimal and will only increase when construction is commercial.

4. HIGH TEMPERATURE GAS COOLED REACTORS

Country	Financing	Institutions	Remarks
Belgium	BF 20 million p.a.	Centre d'études nucléaires (CEN), MOL; Industry.	Study of helium technology and of new irradiation means.
France		CEA as part of an international agreement involving Gulf Atomic International and a group of French firms (Groupement pour l'étude des réacteurs à haute température).	Work related to a possible order for a 1,200 MWe power plant.
"		EDF	Research on fuels, core physics and safety.
Germany	Government financing: 1973: DM.224 million 1974: DM.199 million 1975: DM.175 million 1976: DM.119 million	KFA Jülich in co-operation with nuclear industry. Co-operation with Switzerland in the case of the HHT project.	Three main projects: 1) THTR 300 - High Temperature reactor 280 MWe power plant. Total cost: DM.885 million.
"			2) HHT - Specifications for the construction of a 300 MWe power plant using helium turbines in direct loops. Total cost: DM.251 million. 3) Direct use of nuclear process heat in various industrial processes.
Japan	1974: Yen 367 million	JAERI	Multi-purpose utilisation.
Netherlands		RCN	
Switzerland	FS 15 million until 1975 (5 million p.a.)	Co-operation with Germany through Office de la science et de la recherche. Work carried out by Institut fédéral de recherches en matière de réacteurs with the participation of industry	HHT programme: specifications for construction of a 300 MWe power station with direct-cycle helium turbines. Swiss contribution amounts to 10 per cent of the project.
United Kingdom	1972-73: $12,960,000 1973-74: $9,120,000	UKAEA (Atomic Energy Authority)	Research on problems relating to fuels and to the behaviour of the materials.
"	1972-73: $1,440,000 1973-74: $3,240,000	CEGB	Includes work on "Dragon".
United States	FY 1973: $7.3 million FY 1974: $13.8 million FY 1975: $41.0 million	AEC	R & D work on HTGR in the United States is mainly financed and performed by a private firm, Gulf Atomic International.

N.B. Mention has to be made of the OECD/NEA Dragon project.

5. LIQUID METAL FAST BREEDER REACTORS

Country	Financing	Institutions	Remarks
Belgium	1973-77: FB 2,600 million (15 per cent of the project).		Participation in the German SNR300 project mentioned below.
France	Frs. 400 million p.a. until 1977-78.	CEA (in co-operation with EDF)	i) Demonstration of the system with the Rapsodie reactor and the 300 MWe Phénix power plant. ii) Study of a 1,200 MWe reactor Super-Phénix for a group of French, German and Italian utilities (EDF, RWE, ENEL). iii) R & D work on components, safety and fuel cycles.
Germany	Government financing: 1973: DM.232 million 1974: DM.261 million 1975: DM.318 million 1976: DM.348 million (the German contribution represents 70 per cent of the SNR300 project).	Construction by the international firm INB founded by Interatom, Neratoom and Belgo-nucléaire. The owner and operator will be SBK, the shares of which are held by German, Belgian and Dutch utilities.	Construction of a 300 MWe prototype SNR300. Total cost: DM.1,700 million.
Japan	1972: Yen 19,784 million 1973: Yen 17,558 million 1974: Yen 12,115 million	Power Reactor and Nuclear Fuel Development Corporation.	Construction of a prototype 200-300 MWe FBR scheduled to go critical in 1978, based on an experimental 50MWt reactor started in 1969.
Netherlands	1974: Fl.85 million (15 per cent of the SNR300 project)	RCN and TNO	Participation in the above mentioned German SNR300 project.
Sweden	SK.2.8 million p.a. (includes also work on HTGR).	AB Atomenergi	Study of fuel and material problems, reactor physics and neutron data.
Switzerland	S.Frs.400,000 p.a.	Co-operation with Atomic International Division of North American Rockwell through the Office de la science et de la recherche. Work carried out by the Institut fédéral de recherches en matière de réacteurs.	Evaluation studies.
United Kingdom	1972-73: $78,720,000 1973-74: $76,500,000	UKAEA	Study of alternative designs, development and testing of components, fuel development, fuel reprocessing.
"	1972-73: $1,440,000 1973-74: $3,240,000	CEGB	
United States	FY 1973: $253.7 million FY 1974: $357.3 million FY 1975: $473.4 million	AEC. An important part of the R & D is carried out under contract by industry.	The programme includes the completion in 1980 of a 350 to 400 MWe demonstration plant.

6. OTHER REACTORS

Country	Financing	Institutions	Remarks
Canada	1973-74: Can. $64.3 million	Atomic Energy of Canada Ltd. R & D performed either intramurally or under contract in industry.	R & D activities centred on heavy water reactors, including work on organic coolants, new fuels such as uranium carbide, use of thorium as fertile material.
Japan	1972: Yen 11,539 million 1973: Yen 14,369 million 1974: Yen 12,115 million	Power Reactor and Nuclear Fuel Development Corporation.	Construction of 200MWe prototype ATR, heavy water moderated - boiling light water cooled and fueled with slightly enriched uranium and plutonium mixed oxide. Will go critical in 1976.
Netherlands		KEMA	KSTR (KEMA suspension test reactor).
Switzerland	S.Frs. 2.5 million p.a.	Co-operation with General Atomic Company, with the Nuclear Energy Agency of the OECD and with the European Association "GBRA", through the Office de la science et de la recherche. Work carried out by Institut fédéral de recherches en matière de réacteurs.	Gas cooled fast reactors; installations critical at zero power, studies on thermo-hydraulics of the core and assessment studies.
"	FS 200,000 p.a.	Industry contribution to the work of "GBRA" Association.	
United Kingdom	1972-73: $28,800,000 1973-74: $26,160,000	UKAEA	Includes work on Advanced Gas Cooled Reactor (AGR) SGHWR and supporting research.
United States	FY 1973: $29.5 million FY 1974: $29.0 million FY 1975: $21.4 million	AEC	Light water breeder reactor.

7. SAFETY

Country	Financing	Institutions	Remarks
Australia		Atomic Energy Commission	Work carried out in connection with the licensing and regulation for nuclear energy.
Belgium	BF 4 million p.a.	Centre d'études nucléaires, MOL.	Development of measurement techniques and elaboration of control methods consistent with the non-proliferation treaty.
Denmark	$3 million p.a.	Atomic Energy Commission	
Finland	1974: Fmk. 700,000	Financing by the Ministry of Trade and Industry. Research carried out by the Technical Research Centre of Finland.	Safety of nuclear reactors.
"	1974: Fmk. 650,000	"	Reliability of nuclear reactors.
"	Fmk. 500,000 p.a.	Financing by the Ministry of Trade and Industry. Work carried out in several institutes of the University of Helsinki.	Environmental effects of nuclear energy.
France	Frs. 150 million p.a.	CEA	Co-operation with Germany for the safety of fast breeder reactors. Main work on PWR safety. New programmes on BWR safety.
"	Frs. 40 to 50 million p.a.	CEA	Research on health and environmental effects of radiations.
Germany	Government financing: 1973: DM.53 million 1974: DM.72 million 1975: DM.82 million 1976: DM.96 million		Work on the various aspects of reactor safety.
"	DM.15 million p.a.		Radiation protection and environmental effects of energy transformation.
Japan	1972: Yen 356 million 1973: Yen 1,415 million 1974: Yen 3,914 million	PNC; JAERI; National Institute for Radiological Sciences.	In addition to research concerning the safety of the various types of reactors, a Nuclear Safety Research Reactor is being built. Biological aspects of nuclear safety.
Netherlands		RCN	
Norway	N.Kr. 2.5 million p.a.	Institutt for Atomenergi	Computer Programme for safety analysis of LW Reactors. Analysis of radioactivity dispersion in the environment.
Sweden	SK. 50 million over a period of 4 years.	AB-Atomenergi	Full scale LWR safety experiments at the Marviken reactor station.
"	SK. 12 million for 1973 and 1974.	Delegation for nuclear safety administering a research fund set-up by the utilities.	Projects on short-term safety problems facing industry.
Switzerland	FS 0.5 million p.a.	Work carried out by the Institut fédéral de recherches en matière de réacteurs at the request of the Security Division of nuclear installations.	Studies connected with the licensing of nuclear plants.

N.B. In many cases safety research is part of R & D activities mentioned above on the development of reactors

7. SAFETY (Cont'd)

Country	Financing	Institutions	Remarks
United Kingdom	1972-73: $7.2 million 1973-74: $7.2 million	UKAEA; Department of Energy; CEGB.	The figures shown represent costs of the UKAEA Safety and Reliability Directorate, the Nuclear Inspection Directorate of Department of Energy and the CEGB Nuclear Health and Safety Department. No attempt has been made to dissociate safety expenditure from costs of reactor development.
United States	FY 1973: $38.8 million FY 1974: $48.6 million FY 1975: $61.2 million		

8. RADIO ACTIVE WASTE MANAGEMENT

Country	Financing	Institutions	Remarks
Australia		Atomic Energy Commission	Work on treatment processes for wastes from mining and milling operations.
Belgium	BF 20 million p.a.	Centre d'études nucléaires (CEN), MOL; Industry.	Study and improvement of processing of effluents with a either low or high radioactivity. Development of a new incineration technique to burn plutonium contaminated solid waste.
Austria			Research on storage of radio-active waste.
Canada		Atomic Energy of Canada Limited.	Studies on storage in salt beds and rock formations. Investigation on waste problems in the thorium cycles.
Denmark	US $70,000 p.a.	Atomic Energy Commission	
France	FF 10 million p.a.	CEA	Work on vitrification of solid waste as well as on liquid and gaseous radio-active waste.
Germany	DM.15 million	Three nuclear research centres (GSF, GFK and KFA).	Special attention on the treatment (calcination and vitrification) of highly radio-active waste. Investigation of storage in salt mines.
Japan	1974: Yen 55 million	JAERI	Solidification technology of high or middle level radio-active wastes.
Netherlands		KEMA/RCN	
Norway	N.Kr. 1.7 million	Institutt for Atomenergi	Development of absorption materials and methods for retention of gases. Methods for treatment and disposal of radioactive waste.
Sweden	FY 1973-74: SK 300,000	AB Atomenergi	Main areas: separation of plutonium and other trans-uranium elements, solidification and stabilization of liquid waste, sites for ultimate disposal.
"		KTH (Royal Institut of Technology), Department of Nuclear Chemistry.	Fixation of long lived radio-active waste products in minerals.
"		Chalmers University of Technology, Department of Nuclear Chemistry.	Actinide chemistry.
"		State Power Board	Waste disposal at nuclear power stations.
Switzerland	1974-75: FS 1.5 million (750,000 p.a.)	Group of electrical undertakings with government participation. Research carried out by industry.	Storage of low-activity wastes in geological formations.
"	S.Frs. 25,000 p.a.	Institute fédéral de recherches en matière de réacteurs.	Processing and incineration of low-activity wastes within operating budget.
United Kingdom	1972-73: $3,600,000 1973-74: $3,840,000	UKAEA BNFL; CEGB.	Covers treatment and disposal as well as transportation of waste material.

8. RADIO ACTIVE WASTE MANAGEMENT (Cont'd)

Country	Financing	Institutions	Remarks
United States	FY 1972: $2.5 million FY 1973: $4.2 million FY 1974: $7.8 million FY 1975: $13 million FY 1976: $20.2 million	AEC	Includes programmes on: i) High level waste; Retrievable surface storage, disposal in bedded salt, long range concepts evaluation, solidification, partitioning, cladding hulls; ii) Airborne radioactive effluent; iii) Decontamination methods; iv) Transportation methods and packaging.

9. UNDERLYING RESEARCH

Country	Financing	Institutions	Remarks
Australia		Atomic Energy Commission	The purpose is to build up and maintain competence on those critical aspects of nuclear systems which influence safety and performance. The research covers reactor physics, heat transfer and fluid flow, chemical aspects, materials behaviour, instrumentation and control.
Austria			Studies on neutrophysics, heat transfer, fuel cycles, sodium and helium cycles.
Belgium	BF 175 million p.a.	CEN, MOL; Institut des radio-éléments, Universities.	Solid state and neutron physics. Radiobiology. Exploratory research.
Finland	1974: Fmk. 500,000	Financed partly by the Ministry of Trade and Industry and partly by industry. Research performed by the Technical Research Centre of Finland.	Research on control and operation techniques in nuclear power plants.
"	1974: Fmk. 1.5 million About Fmk. 1 million p.a. in future years.	"	Study on nuclear reactor materials.
Netherlands		RCN; Universities.	Nuclear materials testing.
Sweden	1973-74: SK 0.619 million 1974-75: SK 1.410 million 1975-76: SK 1.090 million	Laboratory of the Research Councils at Studsvik.	Production and study of short lived nuclides.
"		KTH (Royal Institut of Technology), Department of Nuclear Chemistry.	Mechanisms of chemical action of ionising radiation.
"	SK 400,000 p.a.	Chalmers University of Technology, Department of Reactor Physics.	Neutron diffusion and slowing down.
Switzerland		Work carried out by the Institut fédéral de recherches en matière de réacteurs on the installations belonging to the Institut suisse de recherches nucléaires in co-operation with universities.	Setting up a catalogue of nuclear reactions triggered by muonic beams on target elements and of effective cross-section data as a function of incident energy.
United Kingdom	1972-73: $17,520,000 1973-74: $17,280,000	UKAEA	Research on radiation effects, nuclear physics, reactor physics and electronics, materials science, chemistry, neutron beam studies.
United States	$800,000 p.a. Total cost: $10-15 million	National Bureau of Standards	Development of standard reference neutron fields and standard reference neutron sources.

10. THERMONUCLEAR FUSION

Country	Financing	Institutions	Remarks
Belgium	BF 65 million p.a.	CEN, MOL; Universities.	Study of plasmas and magnetic confinement.
Canada	1973-74: Can. $60,000	Supported by the Ministry of State for Science and Technology and the Atomic Energy Control Board.	Feasibility study for a Canadian programme on controlled thermonuclear fusion.
Denmark	Dmk. 600,000 p.a.		
France	1974: Frs. 67 million (partly financed by EURATOM).	CEA. The programme is part of a joint EURATOM programme.	Main emphasis on the creation, confinement and heating of plasmas with magnetic machines such as Tokamak or other machines under construction.
Germany	1973: DM.59 million 1974: DM.68 million 1975: DM.77 million 1976: DM.90 million	Research centres of Garching and Jülich. The programme is part of the joint EURATOM programme.	Magnetic confinement.
Ireland	1972: £4,000 1973: £4,000 1974: £4,000	Department of Electrical Engineering, University College, Cork. Co-operation with UKAEA and French CEA.	Submillimeter diagnostics of ionised plasmas.
Japan	1972: Yen 654 million 1973: Yen 624 million 1974: Yen 991 million	JAERI; Institute of Physical and Chemical Research; Universities.	Magnetic confinement.
Netherlands		FOM/RCN	Fusion reactors.
Sweden	1973-74: SK.1,328,000	KTH (Royal Institute of Technology), Department of Plasma Physics.	Plasma confinement in internal ring systems.
"		Chalmers University of Technology, Department of Nuclear Chemistry.	Chemical problems in fusion technology.
Switzerland	SF 4.7 million p.a.	Centre de recherches en physique des plasmas of the Ecole polytechnique fédérale de Lausanne and Institut de physique of the University of Fribourg.	Plasma physics, confinement and dynamic stabilization of plasmas.
United Kingdom	1972-73: $9.6 million 1973-74: $7.2 million	UKAEA	Work on a large new experimental Tokamak type assembly. This work is part of the joint EURATOM programme.
United States	FY 1972: $33.2 million FY 1973: $39.7 million FY 1974: $57 million	AEC	Magnetic confinement.
" "	FY 1972: $19.4 million FY 1973: $34.5 million FY 1974: $44.1 million	"	Laser fusion.
" "	FY 1974: $0.4 million	DOD	Pulse controlled thermonuclear fusion through electromagnetic implosion of conducting cylinders to obtain multimegagauss fields by flux compression.

11. NUCLEAR SHIP PROPULSION

Country	Financing	Institutions	Remarks
Germany	Government financing 1973: DM.28 million 1974: DM.22 million 1975: DM.29 million 1976: DM.32 million	GKSS	Continuation of research related to the OTTO HAHN merchant vessel. Study of the complete specification for the design and construction of a larger nuclear container vessel of 80,000 shaft h.p.
Japan	1972: Yen 1,724 million 1973: Yen 1,324 million 1974: Yen 1,495 million	Japan Nuclear Ship Development Agency	Research and operation related to nuclear powered ship.
United Kingdom			Under consideration by the Nuclear Ship Steering Group is the possibility of the United Kingdom embarking on nuclear ship work.
United States	$14 million over five years	Maritime Administration	Criteria for design, construction and inspection. Requirements for support facilities, manning, training and licensing. Requirements for port entry, insurance and indemnification.

N.B. This subject also belongs to energy utilisation

1. SOLAR ENERGY

Country	Financing	Institutions	Remarks
Australia	Australian $372,000 p.a.	CSIRO (Commonwealth Scientific and Research Organisation)	Study of methods for collection and use of solar energy and development of appropriate equipment.
Belgium	BF 2 million p.a.	Katholieke Universiteit Leuven (Catholic University of Louvain)	Study of solar batteries and heat transformers.
Denmark	US $30,000		The research programme includes heating and cooling of buildings and solar central stations.
France		CEA (Commissariat à l'énergie atomique) in co-operation with industry and CNES (Centre national d'études spatiales).	Research concerns mainly feasibility studies on photovoltaic and thermal conversion.
	1973: Frs. 5.1 million 1974: Frs. 7.9 million	ANVAR (Agence nationale pour la valorisation de la recherche) and EDF	Development of pilot air-conditioning installations in house units.
		CNRS (Centre national de la recherche scientifique)	Scientific experiments on heat captation by concentration.
		DGRST (Délégation générale à la recherche scientifique et technique)	Terrestrial application of photovoltaic conversion.
			Conversion of solar energy into mechanical energy.
Germany	1974: $1 million 1975: $3 million	EURATOM (ISPRA Joint Research Centre); University of Stuttgart; Industry.	Main emphasis on R & D for heating and cooling of buildings.
Japan	FY 1974: Yen 873 million	MITI (Ministry of International Trade and Industry)	As part of the "Sunshine Project", research on solar energy is scheduled to continue until the year 2000.
			Programme includes: solar power thermal systems, solar cells, thermionic converter, residential heating and cooling, solar furnace.
Netherlands			A study on the use of solar energy for water and space heating is ready to start.
Sweden	FY 1973-1974: S.Kr. 30,000	Natural Science Research Council. Research is carried out at the Royal Institute of Technology.	Hydrogen bond and chemical processes causing photochromatic phenomena.
Switzerland	FS 100,000 p.a.	Office de l'économie énergétique; Office de la science et de la recherche; industry. Research carried out at the Ecole polytechnique fédérale of Lausanne and at Institut suisse de météorologie.	Evaluation studies and meteorological observations.
		Institut Battelle, Geneva	High temperature converters and photoelectric cells.
United Kingdom	1972-73: (small expenditures)	Department of the Environment;	Solar heating.

1. SOLAR ENERGY (Cont'd)

Country	Financing	Institutions	Remarks
United Kingdom	1973-74: $768,000	Some universities	Basic research. Expenditure does not include contribution of EURATOM programme of ISPRA.
United States	FY 1973: $1 million	NASA	Research programme covers: heating and cooling of buildings, power generation, clean fuels production.
	FY 1971: $540,000 FY 1972: $100,000 FY 1973: $483,000 FY 1974: $7,350,000	NSF " NSF/NASA/HUD NSF/NASA/HUD/DOD	Heating and cooling of buildings. " " " " " "
	FY 1971: $60,000 FY 1972: $550,000 FY 1973: $1,430,000 FY 1974: $2,420,000	NSF " " NSF/NASA	Solar thermal energy conversion. " " " " " "
	FY 1972: $409,000 FY 1973: $924,000 FY 1974: $2,579,000	NSF/NASA " "	Photovoltaic conversion programme " " " "
	FY 1972: $2 million FY 1973: $1.7 million FY 1974: $1.8 million	DOD " "	Financing includes programme on solar and geothermal energy. Solar energy programme is directed towards the improvement of efficiency and useful life of solar cells.
	FY 1974: $100,000	Bureau of Mines (Department of Interior)	Materials research.

2. GEOTHERMAL ENERGY

Country	Financing	Institutions	Remarks
Austria			Research on geothermal energy is at present carried out in the framework of geo-scientific and geotechnical programmes.
Canada		Department of Energy, Mines and Resources.	Practical use and exploitation of geothermal energy.
		Earth Physics Branch, Geological Survey of Canada.	Study of hot springs and fumaroles in Western Canada.
		NATO	Participation in the CCMS Task Force of NATO for the improvement of information exchange and non-electrical uses of geothermal energy.
France	1973: Frs. 3.6 million 1974: Frs. 6.2 million	DGRST; CEA; BRGM; IFP (Institut français du pétrole).	Research on geothermal high temperature reservoirs including: expansion and vaporisation mechanisms in rocks, clefts and wells.
			Research on geothermal low temperature reservoirs including: geology, mathematical models of reservoirs evolution, heat exchangers and pumps.
Germany	DM. 0.3 million p.a.	Bundesanstalt für Bodenforschung (Federal Office for Geological Research)	Investigations on prospects of geothermal energy utilisation in volcanic regions.
Iceland	$1.6 million p.a.	National Energy Authority	Exploration of thermal areas by geological, geophysical and geochemical methods, by drilling and hydrological studies. This programme is carried out simultaneously with investigations on hydropower potentials. The financing covers both programmes.
Japan	1972: Yen 90 million 1973: Yen 244 million 1974: Yen 258 million	Research Development Corporation of Japan	Development of technology of steam production used for large scale electric power plant in the Five Year Programme 1972-1976. Total expenditures amount to Yen 759 million.
	FY 1973: Yen 86 million FY 1974: Yen 551 million	MITI	As part of the "Sunshine Project", research on geothermal energy includes: survey of geothermal resources, study of mechanisms of natural steam and hot water reservoirs, development of binary cycles and power plants, large scale test facilities.
United Kingdom			Assessment studies in hand by ETSU (Department of Energy).
United States	FY 1973: $2.5 million FY 1974: $2.8 million	USGS	Research programme: Study of geothermal fields; detection and evaluation of geothermal areas; monitoring of environmental effects; evaluation of geothermal resources.
" "	FY 1974: $4.7 million	AEC	Research includes investigation of dry hot rocks fracturing and developments of improved methods for energy extraction from natural hydrothermal reservoirs.

2. GEOTHERMAL ENERGY (Cont'd)

Country	Financing	Institutions	Remarks
United States	FY 1974: 3.7 million	NSF	Programme objectives: - Resources exploration and assessment; - Environmental, legal and institutional research; - Resources utilisation projects (to obtain operational, technological and economic data for assessment of practicality of commercial production); - Advanced research and technology.
	FY 1974: $100,000	Bureau of Mines	Long-term project for mineral recovery from hot brines.
	FY 1974: $100,000	Bureau of Mines	Investigations on improved materials for handling geothermal brines and steam.
	FY 1972: $2 million FY 1973: $1.7 million FY 1974: $1.8 million	DOD	Investigations on materials and structure in the earth's crust and shallow mantle. Financing includes also programmes on solar energy.

3. ORGANIC MATERIAL AND WASTE

Country	Financing	Institutions	Remarks
		a) Bioconversion	
Canada	1972: Can. $24,000	Manitoba Research Council	Programme objective: maximum utilisation of materials of biological origin as sources of energy or of useful chemicals.
United States	FY 1971: $600,000 FY 1972: $350,000 FY 1973: $680,000 FY 1974: $1,050,000	NSF NSF NSF, NASA NSF, NASA	Programme objective: Economic feasibility of large scale conversion of waste, cultivated organic materials, and water to gaseous, liquid and solid fuels using bio-organism.
		b) Urban Solid Waste	
Denmark	n.a.	n.a.	Utilisation of municipal and industrial waste for energy conversion.
Germany	1974: DM.1 million 1975: DM.3 million	Süd-Westdeutsche Fernwärme	Gasification of solid waste.
Japan	FY 1973: Yen 151 million FY 1974: Yen 402 million	AIST, MITI	Research is carried out by national research institutes and enterprises on a contract basis with MITI. Programme objective: comprehensive resource recovery system from urban waste through conversion (pyrolysis). Study on total system analysis.
Sweden		Royal Institute of Technology	Pyrolysis.
Switzerland	SF 500,000 p.a.	Office de la protection de l'environnement	Research on incineration of domestic and industrial wastes with particular reference to the problem of uniformity of wastes.
United Kingdom	1973-74: $120,000	Department of Industry (Warren Springs Laboratory)	Covers separation of raw refuse into combustible materials for fuel. Funded by Department of Environment.
"	1973-74: $120,000	" "	Covers thermal treatment of waste to produce char and oil by pyrolysis. Also funded by Department of the Environment.

4. OTHERS

Country	Financing	Institutions	Remarks
a) Winds			
Denmark		National Association of Power Stations	Investigations on wind energy carried out in co-operation with industry.
Canada		National Research Council; Defence Research Board.	
Netherlands			A study on the utilisation of winds is ready to start.
Sweden	Sw.Kr. 100,000 p.a.	State Power Board; Board of Technical Development; Telecommunications Administration.	Wind power investigations. Construction of low power units.
Germany	DM 1 million		Studies.
United Kingdom			Prospects being reassessed by ETSU (Department of Energy).
United States	FY 1973: $200,000 FY 1974: $1 million	National Science Foundation	Development of reliable cost-competitive wind energy conversion systems.
b) Tides			
Canada		Nova Scotia Research Foundation	Physical survey in relation to the potential use of the Bay of Fundy tides for generating electric power.
United Kingdom			Prospects being reassessed by CEGB and some universities.
c) Ocean Thermal Gradients			
France		CEA in collaboration with CNEXO (Centre national pour l'exploration des océans).	Feasibility studies concerning the exploitation of ocean thermal gradients.
United States	FY 1972: $ 84,100) FY 1973: $229,200) FY 1974: $700,000)	National Science Foundation	Programme objective: To establish system reliability and economic viability of large scale power plants converting ocean thermal energy into electricity.
d) Waves			
United Kingdom		Department of Energy; Department of the Environment.	Assessment studies by the National Engineering Laboratory on feasibility of project.

1. ELECTRICITY

Country	Financing	Institutions	Remarks
		a) Electricity Generation i) Hydropower	
Iceland		National Energy Authority	Hydrological, topographical and geological investigations of potential hydropower sites. Engineering planning.
Norway		Electricity research institutes in Trondheim. Electric power industry.	
Switzerland	FS 100,000 p.a.	Financing by the Commission fédérale pour l'encouragement des recherches scientifiques. Research carried out by the Ecole polytechnique fédérale of Lausanne.	Research on curvature of water turbine blading.
		ii) Improvement of Power Plants	
Belgium	BF 75 million (approx.) p.a.	Universities; Laboratoire belge de l'industrie électrique.	Various studies connected with the improvement of power plant efficiency.
France		EDF	Regulation and control of power plants. Heat exchange. Auxiliary equipment in power plants. Behaviour of turbo alternators. Studies on the various components of power plants (reliability, corrosion, etc.).
Ireland	1972: £5,000 1973: £5,000 1974: £5,000	Department of Electrical Engineering, University College, Cork.	Transient studies of electrical machines.
Switzerland	FS 100,000 p.a.	Office de la protection de l'environnement. Research carried out by various federal institutes.	Research on composition of fuel oils and combustion gases for conventional power stations. Design of components: burners, boilers, stacks, etc.
United Kingdom	1972-73: $11,280,000 1973-74: $13,500,000	CEGB	Includes work on improved combustion.
		iii) Reduction of Thermal Pollution	
France		EDF	Problems connected with cooling in rivers, lakes and estuaries (cooling capacity, heat dissipation ecology of sites). Dry and wet cooling towers.
Germany			(See Section IV, 2, page 225).
Ireland	£2,000 p.a.	Department of Electrical Engineering, University College, Cork.	Fundamental research on packing ratios in cooling towers.
Japan	FY 1973: Yen 28 million FY 1974: Yen 30 million	MITI	Study of the environmental effect of thermal water from power station. 1st stage (1972-74): evaluation of physical effects. 2nd stage (1975 onwards): evaluation of biological effects.

1. ELECTRICITY

Country	Financing	Institutions	Remarks
		a) Electricity Generation iii) Reduction of Thermal Pollution (Cont'd)	
Sweden		State Power Board	Environmental effects of the outlet of cooling water.
Switzerland	1972-74: FS 400,000 (135,000 p.a.) FS 1 million p.a.	Office de l'économie énergétique; Industry; Institut fédéral de recherches en matière de réacteurs.	Studies on wet and dry cooling towers.
"	1972-74: FS 500,000 (165,000 p.a.)	Office de l'économie énergétique; Institut suisse de méteorologie; Industry.	Climatic effects associated with cooling towers.
United Kingdom	(Not available)		Expenditure on this item is included in Table 8 (Environment), Part III of the Report.
		iv) Turbines and Combined Cycles	
Belgium	BF 4.5 million (Approx.)	Universities; Institut Von Karman.	Research on steam turbines and their components.
France		EDF	Research on steam and gas turbines.
Germany	DM.6 million	STEAG	Combined cycles for utilisation of reject heat (Financed by the Land Nordrhein-Westfalen.)
Netherlands		GASUNIE	Feasibility study on combined cycles.
"		KEMA	Combined cycles.
"		TNO	Higher efficiency of gas turbines.
Sweden		ASEA in collaboration with United Aircraft (United States).	Development of a high efficiency gas turbine which is suitable for peaking power but can also be provided with exhaust gas boilers driving steam turbines.
United States		NASA	Development of O_2/H_2 power peaking turbine to meet hydraulic power need of shuttle-orbiter vehicles. Development of 0.5 - 2.0 kwe gas turbine; gas turbine systems in long life space power generation systems.
		v) MHD	
Japan	Yen 6.4 billion from 1966 to 1975	MITI (Ministry of International Trade and Industry).	Research on long-term operation, heat exchanger, seed recovery, heat-proof materials, superconducting magnets; this work is centred on a 1,000 KW MHD test plan.
Netherlands		Universities	
Sweden	SK 0.4 million p.a.	AB Atomenergi	Exploratory work on open-cycle MHD systems.

1. ELECTRICITY

Country	Financing	Institutions	Remarks
		a) Electricity Generation v) MHD (Cont'd)	
Switzerland	1968-1972: FS 2.1 million	Commission fédérale pour l'encouragement de la recherche scientifique. Government aid temporarily suspended; new possibilities being studied. Research carried out at Institut Battelle, Geneva.	Direct conversion into electricity by closed cycle process.
United States	$0.4 million p.a.	NSF	Theoretical work, small scale experiments, materials research.
" "	FY 1972: $0.3 million FY 1973: $0.3 million FY 1974: $1.5 million	DOD	Basic research into plasmas, research on two phase liquid MHD power systems.
" "	$0.3 million p.a.	Programme financed by the Department of Commerce and the Office of Coal Research, carried out by the National Bureau of Standards.	Research on materials problems associated with MHD generators and big temperature gas turbines.
" "	$1.3 million p.a.	NASA	Demonstration of an MHD channel utilising H_2-O_2 combustion as a source of energy. Initial tests at power levels of 1MW to be completed in 1977.
		vi) Fuel Cells	
Canada		Defence Research Board	Work on portable hydrazine air systems (300w and 60w).
"		Hydro-Quebec	Engineering evaluation of a 70 kw installation.
Germany	Government contribution DM.1-1.5 million p.a.		Research on fuel cell technology for the construction of small power plants situated close to the users.
Netherlands		Universities	
United States		DOD	Development of a family of open cycle fuel cell power plants of 0.5, 1.5, 3 and 5 kw based on thermal cracking of hydrocarbon fuels to generate a hydrogen rich fuel for the phosphoric acid electrolyte fuel cell.
		b) Electricity Storage	
Belgium	BF 10 million (Approx.) p.a.	Universities	Studies of Electrolytes.
France		EDF	Monitoring of the various methods of storage. Design of a 200MW storage plant by compressed air. Studies on hydraulic storage. Research aiming at increasing the specific energy of batteries.
Germany	DM.4.5 million p.a.		Research on batteries and on new methods for storage such as fly wheel storage.

1. ELECTRICITY

Country	Financing	Institutions	Remarks
	b) Electricity Storage (Cont'd)		
Netherlands		Philips	Primary and secondary batteries.
Sweden	Cumulative expenditure up to now: SK13 million	Swedish National Development Company	Iron-air battery.
"		STAL - LAVAL	Air storage gas turbines.
United Kingdom	1972-73 1973-74 Not available	Department of Industry	Development of advanced batteries for buses and light commercial vehicles.
"	1972-73: $180,000 1973-74: $247,000	Department of the Environment and British Rail.	Development of Na/S batteries for rail applications.
"	1972-73: $427,000 1973-74: $286,000	Electricity Council	Research on Na/S batteries.
United States		NSF	Research on batteries, superconducting magnetic storage, mechanical and other forms of storage, in relation to peaking problems of the electric power industry.
" "	FY 1973: $1.5 million FY 1974: $1.3 million	DOD	Lithium-organic electrolyte batteries, Ni/Cd batteries. Solid electrolyte batteries. Zn/air batteries. Ag/Zn batteries.
" "	FY 1973: $0.55 million FY 1974: $1.8 million	AEC	AEC programme includes development of energy storage techniques such as: batteries, hydrogen storage, flywheels, compressed air storage and magnetic energy storage applicable to electric utilities and new energy generating systems (wind, solar, geothermal and fusion). In addition, technologies applicable to transportation are batteries, flywheels and hydride storage.
" "	$1.5 million p.a.	NASA	Ni/Cd and Ag/Zn batteries.
	c) Electricity Transmission		
Belgium	BF 95 million p.a.	Universities, Laboratoire belge de L'industrie électrique.	Research on networks, on conductors and on the automatic control of electricity transport and distribution systems. Research on superconductors.
Austria			Research on high voltages, high intensity currents, AC-DC conversion and superconducting cables.
Canada		National Research Council, Ottawa; Hydro Quebec Institute of Research; Ontario Hydro.	
France		EDF	Research on electrical and mechanical properties of insulators in view of their use in super-conducting cables.

1. ELECTRICITY

Country	Financing	Institutions	Remarks
c) Electricity Transmission (Cont'd)			
Germany	DM.6.5 million p.a.		Transmission lines and very high voltages for DC and AC currents. Cool-end underground transmission lines. Super-conducting lines.
Ireland	1972: £5,000 1973: £5,000 1974: £5,000	Department of Electrical Engineering, University College, Cork.	Study of control systems for high voltage DC transmission schemes.
Japan		MITI	Development of high transmission capacity super-conducting lines; main R & D areas: - electrical behaviour: super-conducting environment; - thyristor valve for AC-DC conversion, super-conducting transformers, cooling systems; - compatibility with present transmission systems.
Netherlands		KEMA	High pressure and gas insulated cables.
Norway	1973: N.Kr.25 million	Electric power industry	High voltages.
Sweden		ASEA	Ultra high voltage transmission. High voltage DC transmission. Innerconnection of power systems.
Switzerland	FS 100,000 p.a.	Commission fédérale pour l'encouragement des recherches scientifiques. Research carried out by the Ecole polytechnique fédérale of Zurich.	Superconducting cables.
United Kingdom	1972-73: $4,968,000 1973-74: $4,632,000	CEGB	Research on ultra high voltage transmission design of various types of cables, including super-conducting cables.
United States	FY 1973: $0.7 million FY 1974: $1 million	AEC	Underground super-conducting cables.
" "		NSF	Compiling and controlling of large systems. Materials for super-conducting cables.

2. HEAT AND DISTRICT HEATING

Country	Financing	Institutions	Remarks
Denmark		National Association of Power Stations	Use of cooling water from power stations.
Finland	1973: FMK 200,000	Financed by the Ministry of Trade and Industry; research performed by the Technical Research Centre of Finland.	Conceptual design of a small size (100-200) low pressure nuclear reactor to be used primarily for the production of district heat.
France		CEA	Use of waste heat from power plants.
Germany	DM.4 million p.a.	Arbeitsgemeinschaft Fernwärme; Kraftanlagen, Heidelberg, STEAG, EVO, BEWAG, WIBERA.	Development of a heat map; feasibility study for district heating.
Netherlands		Philips	Heatpipes.
"		Universities	Heat accumulation and heatpipes.
Sweden		STAL-LAVAL	Development of steam turbines for district heating stations.
"	Preliminary experiment will cost S.Kr. 240,000	Stockholm Energy Works	Transmission of hot water through rock tunnels at long distances.
"		Atomenergi in co-operation with utilities	Long distance water distribution system for district heating especially by low temperature water.
Switzerland	1969-74: FS 400,000 (80,000 p.a.)	Office de l'économie énergétique. Research carried out by industry.	Studies on district heating and on gas and electrical heating.
United Kingdom	1972-73: $240,000 1973-74: $720,000	CEGB Some Universities	Assessment studies on utilisation of waste heat from power stations.
United States	FY 1974: $0.1 million	DOD	Use of waste heat from power plants.
" "		Department of Housing and Urban Development.	Uses of waste heat from power plants for urban heating and cooling. Investigation of the use of small power plants in cities for "total energy", including district heating.

3. OTHER ENERGY CARRIERS

Country	Financing	Institutions	Remarks
France		Gaz de France and CEA	Production of hydrogen through thermochemical dissociation of water in high temperature nuclear reactors. Storage of hydrogen in hydrides.
Germany	DM.1 million		Use of process heat for hydrogen production and hydrocarbon production ("Adam und Eva-Konzept").
Japan	FY 1974: Yen 332 million	MITI	Hydrogen production by electrolysis or thermochemical and thermal decomposition techniques. Transportation and storage of hydrogen. Hydrogen liquefaction. Applications of hydrogen. This activity is part of the "Sunshine Project".
Netherlands		DSM	Feasibility studies on the use of methanol as a fuel for automotive power systems and for fuel cells.
"		Philips	Hydrogen storage in intermetallic compounds.
Switzerland		Institut Battelle, Geneva.	Research on hydrogen storage.
United Kingdom	1973-74: $48,000	AERE (Harwell), ETSU (Department of Energy).	Assessment studies by ETSU and a small amount of experimental work by AERE. Figures do not include United Kingdom contribution to ISPRA work on hydrogen.
United States		NSF	Investigation of the transport and storage of hydrogen.
" "		AEC	Feasibility study of closed cycle thermochemical water cracking for hydrogen production. Hydrogen storage in metal hydrides.
" "	$200,000 p.a.	Bureau of Mines	Investigations of rare earth compounds and alloys for the low pressure storage of hydrogen.

1. RESIDENTIAL & COMMERCIAL SECTOR

Country	Financing	Institutions	Remarks
Australia	Austr. $255,000 p.a.	CSIRO	Thermal design of buildings.
Austria			Thermal insulation of buildings.
Finland	Total costs Fmk 2.4 million	SITRA (foundation controlled by Parliament); financing by several government agencies.	Heat economy in buildings.
France		EDF	All electrical buildings.
"		IFP, CSTB (Centre scientifique et technique du bâtiment.	Improvement of oil fueled domestic heating.
Germany	DM.6 million p.a.		Improvement of the thermal insulation of buildings. Household machines.
Netherlands		DSM; GASUNIE	Thermal insulation of building.
"		Philips	Higher efficiency of gas discharge lamps. Heat pumps and cooling machines.
"		TNO	Heating and cooling of buildings. Higher efficiency of domestic gas heaters.
"		"	Application of heatpump for space heating.
"		VEG	Development of gas burners. Interchangeability of gases.
Sweden		Building Research Council	Various aspects of energy economy in building.
"	1973-74: SK 50,000	Department of Building Technology, Lund University.	Forced convection - thermal performance of insulated building elements as influenced by wind and workmanship.
"		Department of Building Technology, KTH.	Non stationary heat flow in buildings.
"		State Power Board	Heat pump experiments.
Switzerland	1972-74: FS 400,000 (135,000 p.a.)	Office de la protection de l'environnement. Research carried out by the Laboratoire fédéral d'essais des matériaux.	Thermal insulation of buildings.
United Kingdom	1972-73: (Not known) 1973-74: $7,850,000	Department of the Environment, Nationalised Industries Coke Research Association.	Expenditure relates to work on saving and conserving energy by the various bodies mentioned.
United States	FY 1974: $0.1 million	DOD	Performances of building materials and methods.
" "		Department of Housing and Urban Development	Minimisation of energy consumption in homes.
" "	$400,000 p.a.	National Bureau of Standards	Appliance labelling for efficiency of energy utilisation.
" "	$1.9 million p.a.	"	Energy conservation in buildings.

1. RESIDENTIAL & COMMERCIAL SECTOR (Cont'd)

Country	Financing	Institutions	Remarks
United States		Office of Energy Conservation (Department of the Interior)	Demonstration of energy conserving in mobile homes and office buildings. Energy efficiency in military housing. Computer management of heating, ventilating and air conditioning in buildings. Field measurements of actual energy efficiencies in buildings. Efficiencies of ignitors in place of pilot light.

2. INDUSTRY AND AGRICULTURE

Country	Financing	Institutions	Remarks
Belgium		Universities; Laboratoire belge de L'industrie électrique; Centre de recherches métallurgiques.	Research aiming at the reduction of energy consumption in industrial processes.
Canada	FY 1972-73: Can. $230,000 FY 1973-74: Can. $250,000	Department of Energy, Mines and Resources.	Research aiming at improving the efficiency and lowering the cost of pyrometallurgical processes for treating ores.
France		EDF	Investigations of new applications for electricity: e.g. drying in agriculture, or furnaces in steelmaking.
"		GDF	Improvement of natural gas utilisation in large industrial furnaces.
Japan		Japan Heat Management Association (under the guidance of MITI)	Rational utilisation of fuels and heat obtained from fuels in factories.
"	July 1973-March 1979: Yen 7.7 billion	MITI	Direct steelmaking process using high temperature reducing gas.
Netherlands		International Flame Research Foundation	Optimisation of industrial flames and furnaces.
"		TNO	Energy requirements in process industry.
United Kingdom	1972-73: $1,870,000 1973-74: $1,692,000	Electricity Council	Utilisation of electricity in heat, food, chemical and metallurgical industries.
"	1972-73: $3.6 million 1973-74: p.a. for each year	Various organisations	Covers estimated expenditure by big firms like ICI, British Steel and also NIFES for their work on fuel efficiency. Estimate provided by NIFES.
"	1972-73: $550,000 1973-74: $613,000	Ministry of Agriculture Fisheries & Food (MAFF)	R & D to improve energy conversion in tractors and also get most efficient use of agricultural equipment.
United States		Office of Energy Conservation	Investigations for savings in several major energy consuming industries (steel, paper, oil refining, glass, aluminium, rubber, plastics). Improvements in the use of steam in industry.
" "	FY 1973: $772,000 FY 1974: $425,000	AEC	In situ chemical mining of primary sulfide ore, e.g. copper.
" "	$500,000 p.a.	Bureau of Mines	Alternative lower energy consuming mineral processing and substitutes.
" "	$600,000 p.a.	" " "	Energy conservation through recycling of metallic portions of urban refuse.
" "	$500,000 p.a.	" " "	Improved application of in situ leaching of mineral deposits and mining wastes.
" "	$200,000 p.a.	" " "	Use of abundant fuels and reductants in iron ore preparation.
" "	$300,000 p.a.	" " "	Reduction of energy use in the recycling of secondary metals.

3. TRANSPORTATION

Country	Financing	Institutions	Remarks
France		EDF	Research on electric vehicles.
"		IFP	Improvement of motors.
"		I.R.T. (Institut de recherche des transports)	Research on urban vehicles.
Germany	DM.60 million p.a.		New traffic systems.
Japan	FY 1973: Yen 1,366 million FY 1974: Yen 1,130 million	MITI	Research on electric vehicles.
Netherlands		Philips	Petrol engine as a system.
"		"	Non-conventional energy sources for the Stirling engine.
"		TNO	Natural gas as a fuel for diesel engines. Dual fuel automotive systems.
United Kingdom	1972-73: (Not known) 1973-74: $1,920,000	Department of the Environment, Science Research Council (SRC).	Motor vehicle research. Traffic systems. Basic studies sponsored also by SRC. Does not include work by British Rail.
United States	FY 1973: $106.4 million FY 1974: $89 million	DOD	Improvements in propulsion of mobile systems (aircraft, ships and land vehicles).
" "		Department of Transportation and its various administrations	Research on advanced automotive power systems, automotive energy efficiency, transportation systems, gas propulsion, rail propulsion, ground propulsion of aircraft.

Country	Financing	Institutions	Remarks
Canada		Department of Energy, Mines and Resources	Socio-economic programme consisting of collecting and analysing data on all activities and supply-demand conditions in the fields of petroleum, natural gas, uranium and electrical power.
Denmark		Energy Committee	National accounting of energy supply and demand under various assumptions.
France		CEA, EDF.	Simulation and systems analysis techniques.
Germany	DM.5 million	The nuclear research centres of Karlsruhe and Jülich and the Institut für Systemtechnik und Innovationsforschung der Fraunhofer Gesellschaft Karlsruhe.	Systems analysis.
Netherlands		Shell	Development of simulation techniques.
"		Philips	General future research on energy.
Sweden		Chalmers University; Department of Nuclear Chemistry.	Forecasting of future needs for conventional as well as new energy sources.
United Kingdom	1972-73: 1973-74: $2.4 million p.a. for each year	Department of Energy and various other organisations.	Development and application of mathematical modelling to the energy sector of the economy and miscellaneous other economic assessments.
United States	FY 1973: $10.5 million FY 1974: $10.7 million	DOD	Energy systems analysis applied to the energy needs of DOD.
" "		Department of Housing and Urban Development	Research on integrated utility systems.

Annex II

COUNTRY OR REGIONAL
ENERGY BALANCE TABLES(1)

1) These tables are derived from:
Energy Prospects to 1985 -
An assessment of Long-Term Energy Developments and Related Policies, Vol.II, OCDE, Paris 1975.

Australia 1972

Mtoe (10^{13} kcal)	Coal	Oil	Gas	Electricity	Nuclear	Hydro & Geo	Total
Indigenous Supply	43.0	15.9	2.9		-	1.2	63.0
Imported Supply (net)	-16.6	12.5	-			-	-4.1
Stock Changes(1)	-3.0	-0.1	-				-3.1
TOTAL PRIMARY ENERGY	23.4	28.3	2.9		-	1.2	55.8(2)
Electricity Generation	14.0	0.5	0.7	-4.8	-	1.2	11.6
TOTAL FINAL CONSUMPTION	9.4	27.8	2.2	4.8			44.2
Energy Sector & Losses	1.9	2.5	0.5	0.7			5.6
Industry	6.4	5.1	1.1	2.5			15.1
Transportation	-	14.6(3)		0.1			14.7
Residential/ Commercial	1.1	3.7	0.6	1.6			7.0
Non-Energy Oil		1.9					1.9

1) Withdrawals (+), Additions (-)

2) A further 2 Mtoe of wood and bagasse are used for energy purposes

3) Includes 9.3 gasoline; 2.5 ships' bunkers

Canada 1972

Mtoe (10^{13} kcal)	Coal	Oil	Gas	Electri-city	Nuclear	Hydro & Geo	Total
Indigenous Supply	11.7	95.8	56.9		1.9	19.1	185.6
Imported Supply(net)	7.0	-12.8	-24.5	(-0.7)		-	-30.3
Stock Changes(1)	-0.9	0.7					-0.2
TOTAL PRIMARY ENERGY	17.8	83.9	32.4	-	1.9	19.1	155.1
Electricity Generation	9.9	2.6	2.6	-20.3	1.9	19.1	15.8
TOTAL FINAL CONSUMPTION	7.9	81.3	29.8	20.3			139.3
Energy Sector & Losses	1.8	6.2	3.5	2.1			13.6
Industry	5.6	10.0	13.1	9.2			37.9
Transportation	-	34.3(2)	-	-			34.3
Residential/ Commercial	0.5	25.1	13.2	9.0			47.8
Non-Energy Oil		5.7					5.7

1) Withdrawals (+), Additions (-)

2) Includes 24.8 gasoline; 2.6 ships' bunkers

E.E.C. 1972

Mtoe (10^{13} kcal)	Coal	Oil	Gas	Electricity	Nuclear	Hydro & Geo	Total
Indigenous Supply	214.5	11.9	110.7		13.9	12.2	363.2
Imported Supply(net)	21.0	580.8	3.2				605.0
Stock Changes(1)	-4.8	-4.1	-0.6				-9.5
TOTAL PRIMARY ENERGY	230.7	588.6	113.3		13.9	12.2	958.7
Electricity Generation	110.4	68.9	21.7	-83.1	13.9	12.2	144.0
TOTAL FINAL CONSUMPTION	120.3	519.7	91.6	83.1			814.7
Energy Sector & Losses	15.4	31.1	4.7	10.0			61.4
Industry	62.5	125.1	47.3	38.8			273.7
Transportation	1.2	164.4(2)	0.1	2.0			167.9
Residential/ Commercial	41.0	146.5	39.5	32.3			259.3
Non-Energy Oil		52.4					52.4

1) Withdrawals (+), Additions(-)

2) Includes 76.0 gasoline; 39.1 ships' bunkers

Japan 1972

Mtoe (10^{13} kcal)	Coal	Oil	Gas	Electri-city	Nuclear	Hydro & Geo	Total
Indigenous Supply	18.8	0.7	2.7		2.3	9.4	33.9
Imported Supply (net)	38.9	247.5	1.3				287.7
Stock Changes(1)	-0.5	-2.8	-				-3.3
TOTAL PRIMARY ENERGY	57.2	245.4	4.0		2.3	9.4	318.3
Electricity Generation	5.7	65.5	1.5	-37.9	2.3	9.4	46.5
Manufactured Gas	4.5	3.0	-5.1	-			2.4
TOTAL FINAL CONSUMPTION	47.0	176.9	7.9	37.9			269.4
Energy Sector & Losses	6.0	10.3	0.8	3.5			20.6
Industry	39.7	52.1	2.0	24.7			118.5
Transportation	0.4	48.4(2)	-	1.8			50.6
Residential/ Commercial	0.9	31.7	4.9	7.9			45.4
Non-Energy Oil		34.3					34.3

1) Withdrawals (+), Additions (-)

2) Includes 20.0 gasoline; 11.9 ships' bunkers

OECD Europe 1972

Mtoe (10^{13} kcal)	Coal	Oil	Gas	Electricity	Nuclear	Hydro & Geo	Total
Indigenous Supply	230.1	19.8	112.6		16.5	35.9	414.9
Imported Supply (net)	31.4	718.4	5.5				755.3
Stock Changes(1)	-4.6	-7.5	-0.6				-12.7
TOTAL PRIMARY ENERGY	256.9	730.7	117.5		16.5	35.9	1157.5
Electricity Generation	118.5	81.5	22.7	-111.0	16.5	35.9	164.1
TOTAL FINAL CONSUMPTION	138.4	649.2	94.8	111.0			993.4
Energy Sector & Losses	18.3	35.2	5.2	13.5			72.2
Industry	73.7	161.1	49.0	53.2			337.0
Transportation	2.1	207.4(2)	0.1	2.7			212.3
Residential/ Commercial	44.3	186.7	40.5	41.6			313.1
Non-Energy Oil		58.8					58.8

1) Withdrawals (+), Additions (-)

2) Includes 93.7 gasoline; 45.6 ships' bunkers

New Zealand 1972

Mtoe (10^{13} kcal)	Coal	Oil	Gas	Electricity	Nuclear	Hydro & Geo	Total
Indigenous Supply	1.3	0.1	0.3			1.6	3.3
Imported Supply (net)	~0(1)	4.1	-				4.1
TOTAL PRIMARY ENERGY	1.3	4.2	0.3			1.6	7.4
Electricity Generation	0.3	0.2	0.1	-1.5		1.6	0.7
TOTAL FINAL CONSUMPTION	1.0	4.0	0.2	1.5			6.7
Energy Sector & Losses	~0(1)	0.3	0.1	0.2			0.6
Industry	0.6	1.1	0.1	0.5			2.3
Transportation		2.3		~0(1)			2.3
Residential/ Commercial	0.4	0.2	~0(1)	0.8			1.4
Non-Energy Oil		0.1					0.1

1) Less than 0.05 Mtoe

United States 1972

Mtoe(10^{13} kcal)	Coal	Oil	Gas	Electricity	Nuclear	Hydro & Geo	Total
Indigenous Supply	365.4	560.3	571.6		14.1	29.7	1541.1
Imported Supply (net)	-38.8	254.6	24.8		-	-	240.6
Stock Changes(1)	-13.4	9.9	-9.0				-12.5
TOTAL PRIMARY ENERGY	313.2	824.8	587.4		14.1	29.7	1769.2
Electricity Generation	191.0	80.8	104.8	-169.8	14.1	29.7	250.6
TOTAL FINAL CONSUMPTION	122.2	744.0	482.6	169.8			1518.6
Energy Sector & Losses	15.0	35.7	28.1	22.2			101.0
Industry	97.3	50.8	262.2	64.5			474.8
Transportation	0.2	430.9(2)	-	0.4			431.5
Residential/ Commercial	9.7	140.9	192.3	82.7			425.6
Non-Energy Oil		85.7					85.7

1) Withdrawals (+), Additions (-)

2) Includes 308.6 gasoline; 7.7 ships' bunkers

Annex III

PANEL OF EXPERTS

MEETING ON THE SCIENTIFIC AND TECHNOLOGICAL ASPECTS OF ENERGY PROBLEMS

OECD, 21st/22nd March, 1974

Pr. Umberto COLOMBO
Chairman of the
Committee for Scientific
and Technological Policy

Pr. Pierre AIGRAIN
Visiting Professor,
M.I.T., Cambridge, Mass.

Pr. I.C. BUPP
Harvard University, Cambridge, Mass.

M. Bernard DELAPALME
ELF-ERAP, Paris

Dr. J.C. DERIAN
Center for Policy Alternatives,
M.I.T., Cambridge, Mass.

Ir. Edwin C. GOLDMAN
Shell International, The Hague

Pr. Dr. Wolf HAFELE
International Institute for
Applied Systems Analysis, Vienna

Dr. F. HAGEMAN
Shell International, The Hague

Dr. R. LOFTNESS
Electric Power Research Institute, Palo Alto

Dr. H.D. SCHILLING
Steinkohlenbergbauverein, Essen

Pr. Chauncey STARR
Electric Power Research Institute, Palo Alto

Dr. M.A. STEINBERG
Lockheed Aircraft Corporation, Burbank

Pr. Carroll WILSON
M.I.T., Cambridge, Mass.

Annex IV

CONVERSION FACTORS FOR UNITS OF ENERGY

1 cal = 4.1868 joules
1 BTU = 1,055.06 "
= 252 cal

---

1 toe = 10^7 Kcal
= 4.1868×10^{10} joules

---

1 barrel = 0.1416 toe
= 0.5928×10^{10} joules
1 barrel per day = 51.7 toe per year

---

1 Kwh = 3,600,000 joules

Annex V

CONVERSION RATES OF NATIONAL CURRENCIES IN US$

COUNTRY	1972	1973 (and 1974)
Australia	1.191	1.40
Austria	0.043234	0.0481
Belgium	0.022713	0.0235
Canada	1.009887	1.009
Denmark	0.143778	0.1600
Finland	0.240618	0.249
France	0.198289	0.2170
Germany	0.31371	0.3438
Greece	0.03333	0.03333
Iceland	0.01439	0.01439
Ireland	2.501369	2.421
Italy	0.001714	0.001754
Japan	0.003300	0.003714
Netherlands	0.311542	0.343411
Norway	0.15182	0.1654
Portugal	0.03689	0.0397
Spain	0.01556	0.0168
Sweden	0.210204	0.2231
Switzerland	0.261925	0.30597
Turkey	0.071428	0.071428
United Kingdom	2.501369	2.421
United States	1	1

OECD SALES AGENTS
DEPOSITAIRES DES PUBLICATIONS DE L'OCDE

ARGENTINA - ARGENTINE
Carlos Hirsch S.R.L.,
Florida 165, BUENOS-AIRES.
☎ 33-1787-2391 Y 30-7122

AUSTRALIA - AUSTRALIE
B.C.N. Agencies Pty, Ltd.,
161 Sturt St., South MELBOURNE, Vic. 3205.
☎ 69.7601
658 Pittwater Road, BROOKVALE NSW 2100.
☎ 938 2267

AUSTRIA - AUTRICHE
Gerold and Co., Graben 31, WIEN 1.
☎ 52.22.35

BELGIUM - BELGIQUE
Librairie des Sciences
Coudenberg 76-78, B 1000 BRUXELLES 1.
☎ 13.37.36/12.05.60

BRAZIL - BRESIL
Mestre Jou S.A., Rua Guaipá 518,
Caixa Postal 24090, 05089 SAO PAULO 10.
☎ 256-2746/262-1609
Rua Senador Dantas 19 s/205-6, RIO DE JANEIRO GB. ☎ 232-07. 32

CANADA
Information Canada
171 Slater, OTTAWA. KIA 0S9.
☎ (613) 992-9738

DENMARK - DANEMARK
Munksgaards Boghandel
Nørregade 6, 1165 KØBENHAVN K.
☎ (01) 12 69 70

FINLAND - FINLANDE
Akateeminen Kirjakauppa
Keskuskatu 1, 00100 HELSINKI 10. ☎ 625.901

FRANCE
Bureau des Publications de l'OCDE
2 rue André-Pascal, 75775 PARIS CEDEX 16.
☎ 524.81.67
Principaux correspondants :
13602 AIX-EN-PROVENCE : Librairie de l'Université. ☎ 26.18.08
38000 GRENOBLE : B. Arthaud. ☎ 87.25.11
31000 TOULOUSE : Privat. ☎ 21.09.26

GERMANY - ALLEMAGNE
Verlag Weltarchiv G.m.b.H.
D 2000 HAMBURG 36, Neuer Jungfernstieg 21
☎ 040-35-62-501

GREECE - GRECE
Librairie Kauffmann, 28 rue du Stade,
ATHENES 132. ☎ 322.21.60

ICELAND - ISLANDE
Snaebjörn Jónsson and Co., h.f.,
Hafnarstræti 4 and 9, P.O.B. 1131,
REYKJAVIK. ☎ 13133/14281/11936

INDIA - INDE
Oxford Book and Stationery Co. :
NEW DELHI, Scindia House. ☎ 47388
CALCUTTA, 17 Park Street. ☎ 24083

IRELAND - IRLANDE
Eason and Son, 40 Lower O'Connell Street,
P.O.B. 42, DUBLIN 1. ☎ 01-41161

ISRAEL
Emanuel Brown :
35 Allenby Road, TEL AVIV. ☎ 51049/54082
also at :
9, Shlomzion Hamalka Street, JERUSALEM.
☎ 234807
48 Nahlath Benjamin Street, TEL AVIV.
☎ 53276

ITALY - ITALIE
Libreria Commissionaria Sansoni :
Via Lamarmora 45, 50121 FIRENZE. ☎ 579751
Via Bartolini 29, 20155 MILANO. ☎ 365083
Sous-dépositaires :
Editrice e Libreria Herder,
Piazza Montecitorio 120, 00186 ROMA.
☎ 674628
Libreria Hoepli, Via Hoepli 5, 20121 MILANO.
☎ 865446
Libreria Lattes, Via Garibaldi 3, 10122 TORINO.
☎ 519274
La diffusione delle edizioni OCDE è inoltre assicurata dalle migliori librerie nelle città più importanti.

JAPAN - JAPON
OECD Publications Centre,
Akasaka Park Building,
2-3-4 Akasaka,
Minato-ku
TOKYO 107. ☎ 586-2016
Maruzen Company Ltd.,
6 Tori-Nichome Nihonbashi, TOKYO 103,
P.O.B. 5050, Tokyo International 100-31.
☎ 272-7211

LEBANON - LIBAN
Documenta Scientifica/Redico
Edison Building, Bliss Street,
P.O.Box 5641, BEIRUT. ☎ 354429 - 344425

THE NETHERLANDS - PAYS-BAS
W.P. Van Stockum
Buitenhof 36, DEN HAAG. ☎ 070-65.68.08

NEW ZEALAND - NOUVELLE-ZELANDE
The Publications Officer
Government Printing Office
Mulgrave Street (Private Bag)
WELLINGTON. ☎ 46.807
and Government Bookshops at
AUCKLAND (P.O.B. 5344). ☎ 32.919
CHRISTCHURCH (P.O.B. 1721). ☎ 50.331
HAMILTON (P.O.B. 857). ☎ 80.103
DUNEDIN (P.O.B. 1104). ☎ 78.294

NORWAY - NORVEGE
Johan Grundt Tanums Bokhandel,
Karl Johansgate 41/43, OSLO 1. ☎ 02-332980

PAKISTAN
Mirza Book Agency, 65 Shahrah Quaid-E-Azam,
LAHORE 3. ☎ 66839

PHILIPPINES
R.M. Garcia Publishing House,
903 Quezon Blvd. Ext., QUEZON CITY,
P.O. Box 1860 - MANILA. ☎ 99.98.47

PORTUGAL
Livraria Portugal,
Rua do Carmo 70-74. LISBOA 2. ☎ 360582/3

SPAIN - ESPAGNE
Libreria Mundi Prensa
Castelló 37, MADRID-1. ☎ 275.46.55
Libreria Bastinos
Pelayo, 52, BARCELONA 1. ☎ 222.06.00

SWEDEN - SUEDE
Fritzes Kungl. Hovbokhandel,
Fredsgatan 2, 11152 STOCKHOLM 16.
☎ 08/23 89 00

SWITZERLAND - SUISSE
Librairie Payot, 6 rue Grenus, 1211 GENEVE 11.
☎ 022-31.89.50

TAIWAN
Books and Scientific Supplies Services, Ltd.
P.O.B. 83, TAIPEI.

TURKEY - TURQUIE
Librairie Hachette,
469 Istiklal Caddesi,
Beyoglu, ISTANBUL, ☎ 44.94.70
et 14 E Ziya Gökalp Caddesi
ANKARA. ☎ 12.10.80

UNITED KINGDOM - ROYAUME-UNI
H.M. Stationery Office, P.O.B. 569, LONDON
SE1 9 NH, ☎ 01-928-6977, Ext. 410
or
49 High Holborn
LONDON WC1V 6HB (personal callers)
Branches at: EDINBURGH, BIRMINGHAM, BRISTOL, MANCHESTER, CARDIFF, BELFAST.

UNITED STATES OF AMERICA
OECD Publications Center, Suite 1207,
1750 Pennsylvania Ave, N.W.
WASHINGTON, D.C. 20006. ☎ (202)298-8755

VENEZUELA
Libreria del Este, Avda. F. Miranda 52,
Edificio Galipán, Aptdo. 60 337, CARACAS 106.
☎ 32 23 01/33 26 04/33 24 73

YUGOSLAVIA - YOUGOSLAVIE
Jugoslovenska Knjiga, Terazije 27, P.O.B. 36,
BEOGRAD. ☎ 621-992

Les commandes provenant de pays où l'OCDE n'a pas encore désigné de dépositaire
peuvent être adressées à :
OCDE, Bureau des Publications, 2 rue André-Pascal, 75775 Paris CEDEX 16
Orders and inquiries from countries where sales agents have not yet been appointed may be sent to
OECD, Publications Office, 2 rue André-Pascal, 75775 Paris CEDEX 16

OECD PUBLICATIONS, 2, rue André-Pascal, 75775 Paris Cedex 16 - No. 34.277 1975

PRINTED IN FRANCE